AF713298

COMPOSITIONAL ABDUCTION AND SCIENTIFIC INTERPRETATION

Abductive reasoning is a form of inference that infers some hypothesis because of what that hypothesis explains. Unlike deductive reasoning, it yields a plausible conclusion but does not definitively verify it. The theory of compositional abduction developed in this book provides a novel theory of confirmation. Kenneth Aizawa uses case studies to analyze how scientists interpret the results of experiments to support compositional hypotheses (i.e., hypotheses about what things are composed of) and suggests that they use a kind of abduction. His theory is offered as an alternative account of scientific reasoning that the logical empiricists would have interpreted as hypothetico-deductive confirmation. It is also an alternative to the Peircean interpretation of the role of abduction in science. It will be of interest to philosophers of science, especially those working on hypothetico-deductive confirmation, Peirce's view of abduction, inference to the best explanation, and the New Mechanism. This title is also available as Open Access on Cambridge Core.

KENNETH AIZAWA is Professor of Philosophy at Rutgers University, Newark. He is coeditor (with Carl Gillett) of *Scientific Composition and Metaphysical Ground* (2016), coauthor (with Fred Adams) of *The Bounds of Cognition* (2008), and author of *The Systematicity Arguments* (2003).

COMPOSITIONAL ABDUCTION AND SCIENTIFIC INTERPRETATION

A Granular Approach

KENNETH AIZAWA

Rutgers University

CAMBRIDGE
UNIVERSITY PRESS

CAMBRIDGE
UNIVERSITY PRESS

Shaftesbury Road, Cambridge CB2 8EA, United Kingdom

One Liberty Plaza, 20th Floor, New York, NY 10006, USA

477 Williamstown Road, Port Melbourne, VIC 3207, Australia

314–321, 3rd Floor, Plot 3, Splendor Forum, Jasola District Centre, New Delhi – 110025, India

103 Penang Road, #05-06/07, Visioncrest Commercial, Singapore 238467

Cambridge University Press is part of Cambridge University Press & Assessment, a department of the University of Cambridge.

We share the University's mission to contribute to society through the pursuit of education, learning and research at the highest international levels of excellence.

www.cambridge.org
Information on this title: www.cambridge.org/9781009435697

DOI: 10.1017/9781009435710

© Kenneth Aizawa 2025

This publication is in copyright. Subject to statutory exception and to the provisions of relevant collective licensing agreements, with the exception of the Creative Commons version the link for which is provided below, no reproduction of any part may take place without the written permission of Cambridge University Press & Assessment.

An online version of this work is published at doi.org/10.1017/9781009435710 under a Creative Commons Open Access license CC-BY-NC 4.0 which permits re-use, distribution and reproduction in any medium for non-commercial purposes providing appropriate credit to the original work is given and any changes made are indicated. To view a copy of this license visit https://creativecommons.org/licenses/by-nc/4.0

When citing this work, please include a reference to the DOI 10.1017/9781009435710

First published 2025

Cover image: *Athena's Conception*, 2024 © Conner St. Claire

A catalogue record for this publication is available from the British Library

A Cataloging-in-Publication data record for this book is available from the Library of Congress

ISBN 978-1-009-43569-7 Hardback

Cambridge University Press & Assessment has no responsibility for the persistence or accuracy of URLs for external or third-party internet websites referred to in this publication and does not guarantee that any content on such websites is, or will remain, accurate or appropriate.

Contents

Figures

Tables

Acknowledgments

Thanks to Mark Couch for feedback on Chapters 1–4. A special thanks to Drew Headley for his role as an outstanding collaborator. His early help with Hodgkin and Huxley's work opened up the possibilities for this project. Thanks also to Drew for feedback on Chapters 1–5. Thanks to Caitlin Mace for comments on some material from this book presented at the 2024 Pacific Division meeting of the American Philosophical Association and to Zoe Drayson for organizing the session of which that was a part. Thanks to Edouard Machery for arranging a Lunch Time Talk at the Center for the Philosophy of Science at the University of Pittsburgh. Thanks to Mousa Mohammadian for comments on Chapter 7 and for helpful conversation at Pitt. Sincere thanks to Hilary Gaskin for seeing the promise in this work where few others did.

My greatest debt is to Angie, the love of my life. Without her love and support I would not have finished.

Introduction

In 1952, Alan Hodgkin and Andrew Huxley argued that sodium ions carry the initial inward current of the action potential. Historians and philosophers of science will want an account of what it is for sodium ions to carry the initial inward current and what Hodgkin and Huxley's arguments were. In recent years, the gist of an answer to the ontological question – what it is for sodium ions to carry the initial inward current – has emerged. Hodgkin and Huxley meant that an influx of sodium ions into the axon noncausally determines the inward current. There are multiple variations of this answer and, of course, challenges to it. Still, the ontological question has been the subject of active debate in the philosophy of science.

In contrast to the ontological question, there has yet to be any serious philosophical effort to address the methodological question of Hodgkin and Huxley's arguments for the sodium hypothesis. What reasons did they offer for believing that sodium carries the initial inward current of the action potential? For a serious effort, one must at least review Hodgkin and Huxley's experimental work. One of these experiments involved depolarizing an axon by 65 mV and then measuring the change in current across the membrane, first in a sodium-containing medium (seawater), then in a sodium-free medium, then again in a sodium-containing medium[1] (see Figure 0.1). When the axon is depolarized in the sodium-free medium, the initial inward current (represented by the initial rise at the left of the curve) disappears.

How did Hodgkin and Huxley interpret the results shown in figure 0.1? Results do not simply speak for themselves. More concretely, how did Hodgkin and Huxley link the manipulation of an axon membrane potential and a measurement of the subsequent axonal currents to unmeasured movements of ions? Part of the story is that Hodgkin and Huxley realized that an influx of sodium after voltage clamping in seawater would explain

[1] Hodgkin and Huxley (1952a, pp. 450–451).

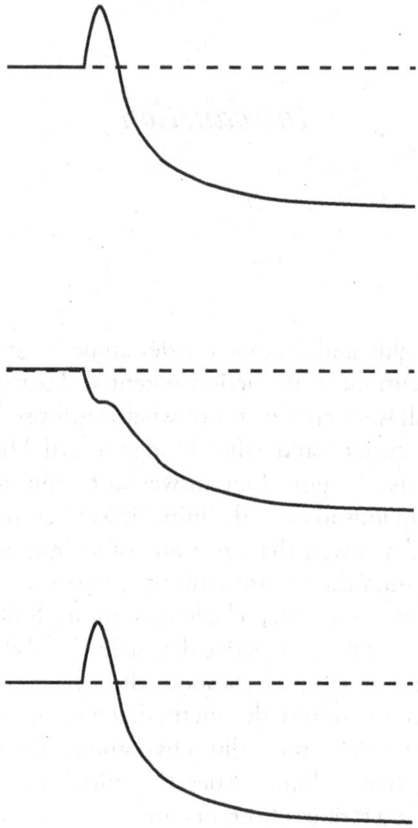

Figure 0.1 Currents in axon no. 15 depolarized by 65 mV in a sodium-containing medium (top panel), sodium-free medium (middle panel), and sodium-containing medium (bottom panel). Redrawn from Hodgkin and Huxley (1952a, p. 451, figure 1).

the currents. Further, they assumed that this explanation provides evidence supporting the hypothesis that sodium carries the initial inward current. They supported the sodium hypothesis based on the idea that this hypothesis would explain their experimental results. In brief, Hodgkin and Huxley were engaged in abductive reasoning.

0.1 Singular Compositional Abduction

The methodological starting point of this book is a specific take on the idea that the history and philosophy of science should be scientifically engaged.

The take is that one important method for a historian and philosopher of science who wishes to understand how scientists confirm hypotheses about noncausal dependence relations is to closely attend to how scientists do this in the experimental literature. Focus on how, in the scientific literature, scientists interpret experimental results for their peers. Focus, for example, on Hodgkin and Huxley's interpretation of the results shown in figure 0.1. This is what William Wimsatt might have accepted as looking at "actual inferences from actual data" (Wimsatt, 1994, p. 207).

I propose that Hodgkin and Huxley interpreted their results using "singular compositional abduction." So, what is singular compositional abduction? Abductive reasoning, in general, is often taken to be a matter of inferring some hypothesis, because that hypothesis explains something.[2] It is also sometimes proposed that abductive inferences may be individuated, at least partly, by reference to the explanations they invoke.[3] Thus, etiological explanations figure into etiological abductions. Compositional explanations figure into compositional abductions. Just so, singular compositional explanations figure into singular compositional abductions.

So, what are singular compositional explanations? They are representations of compositional many-one ontological dependence relations between spatially and temporally localizable particulars. For concreteness, consider further the Hodgkin–Huxley example. A scientific result is a spatially and temporally localized particular. One day, in the late summer of 1949, Hodgkin and Huxley depolarized axon no. 15 by 65 mV and measured the subsequent currents, twice in a sodium-containing solution and once in a sodium-free solution. Figure 0.1 graphically represents three currents – three spatiotemporal particulars – generated in the same lab not too many minutes apart. Hodgkin and Huxley treated these results – each of these particulars – as an explanandum, that is, something to be explained. That which does the explaining – the explanans – in each of the cases is the movement of many sodium ions across the membrane. A singular compositional explanation, in this case, explains an instance of an activity of axon no. 15 in terms of instances of activities of individual sodium ions.[4] The ontological dependence relation is many-one since

[2] See, for example, Peirce (1992), Bird (2005), and Aliseda (2006).

[3] Psillos (2007), Schurz (2008), Schupbach (2017).

[4] The "New Mechanists" in the philosophy of science might take these to be cases of "singular mechanistic constitutive explanations" by embracing the "singular" aspect of the explanation. See, for example, Machamer et al. (2000), Craver (2007), Bechtel and Richardson (2010), and Craver and Darden (2013).

there are many activity instances of sodium ions standing in an ontological dependence relation to an activity instance of axon no. 15.

As brilliant as Hodgkin and Huxley were and as transformative as was their development of their theory of the action potential, among their innovations was not the use of singular compositional abduction. Indeed, singular compositional abduction is, on the one hand, widely used in chemistry, biology, and psychology, but largely unexamined by historians and philosophers of science, on the other. This, despite the widespread commitment among historians and philosophers of science to look closely at actual scientific practice. This, despite the widespread presupposition that scientists make pervasive use of abduction.

Enormous tracts of research in animal behavior, chemistry, child development, generative linguistics, and psychology would not exist were it not for singular compositional abduction. However, consider just one concrete illustration. In a much-discussed *Psychological Review* paper published in 1948, Edward Tolman argued that rats sometimes navigate mazes using a cognitive map, rather than simpler "stimulus-response" strategies.[5] Such maps, he proposed, indicate "routes and paths and environmental relationships, which finally [determine] what responses, if any, the animal will finally release" (Tolman, 1948, p. 192). Consider one of the simplest experiments Tolman discussed, performed by Kenneth Spence and Ronald Lippitt in the Psychological Laboratory at the State University of Iowa sometime between 1939 and 1942.[6] The experiment involved a Y-maze with food in one goal arm and water in the other. During the training period, each of the twenty rats in the experiment was allowed to roam the maze freely four times per day, neither thirsty nor hungry. During the test phase, ten of the rats were hungry, whereas the other ten were thirsty. Unsurprisingly to most contemporary philosophers and psychologists, the thirsty rats were more likely to go to the water, but the hungry rats were more likely to go to the food. Tolman interpreted these results in terms of a cognitive map. He proposed that the thirsty rats had formed a cognitive map of the maze that specified that food was in one arm and water in the other and they wanted water. He also proposed that the hungry rats had formed a cognitive map of the maze that specified that food was in one arm and water in the other and they wanted food. Given manipulations of the rats and measurements of their behavior, Tolman postulated an unseen and unmeasured cognitive map. Why? Because

[5] Tolman (1948).

[6] Tolman cites an unpublished manuscript that subsequently appeared as Spence and Lippitt (1946).

instances of the activity of a cognitive map in those rats would explain the behavior (the instances of the activities) of those rats in the Iowa Psychological Laboratory sometime in 1939–1942.

0.2 Some Philosophical Context

My "science-first" approach to actual inferences from actual results reveals details often unnoticed by philosophers of science. C. S. Peirce, for example, proposed that abductive inference does not support hypotheses, but merely introduces them or suggests them as worthy of further examination by other methods.[7] However, I will argue that Hodgkin and Huxley, among others, tried to use a series of compositional abductive inferences to confirm hypotheses. One might defend Peirce by noting that he could not have foreseen how mid twentieth-century scientists would use abductive reasoning. Nevertheless, many later philosophers have followed Peirce's understanding of the role of abductive reasoning in science.[8]

Gilbert Harman proposed that warranted abductive inference is a matter of inference to the best explanation (IBE), where what determines the best explanation is a matter of, among other things, what is simpler, more plausible, and has broader scope.[9] Further, he mentioned that IBE is used in science. This account has been extremely influential among philosophers of science who study abduction, but Harman says nothing about how IBE might apply, if at all, to local scientific arguments, such as Hodgkin and Huxley's interpretation of their experimental results. Harman was engaged in an epistemological project of trying to show that enumerative induction is merely a special case of IBE. He was, therefore, noncommittal about exactly how his picture might apply, if at all, to specific scientific arguments.

Peter Lipton developed what is widely recognized to be the most advanced version of Harman's idea of IBE and its application to scientific reasoning.[10] Unlike in Harman, the preponderance of the evidence shows that Lipton intended his account of IBE to be applicable in many contexts, including the scientific interpretation of experimental results. In the context of interpreting experiments, Lipton proposed that explanatory questions have a contrastive structure, such as "Why did the women in Clinic 1 have a high rate of mortality from childbed fever, but those in Clinic

[7] Peirce (1992).
[8] For example, Hanson (1960), Magnani (2001), and Gabbay and Woods (2005).
[9] Harman (1965). [10] Lipton (2003).

2 did not?" Further, he proposed that these contrastive questions align with controlled experiments as he illustrates with Ignaz Semmelweis's interpretation of experimental results concerning childbed fever.

Close attention to scientific experiments and arguments also reveals details unexamined in the New Mechanist philosophy of science literature. Although the Hodgkin–Huxley model is something of a fixture in this literature, there has been no attention to Hodgkin and Huxley's use of experimental work to support the sodium hypothesis.[11] Part of the story is that in Hodgkin and Huxley's first experiment, an axon is voltage-clamped and a current is measured. This is an experiment with an intervention at the level of the axon and a measurement at that same level. This is what might be called an "intralevel experiment."[12] (The Tolman example also relied on an intralevel experiment.) Based on this, Hodgkin and Huxley advanced a hypothesis about sodium ion fluxes. New Mechanists have never discussed the possibility that intralevel experiments might be used to confirm relations of noncausal dependence.[13] New Mechanists have never considered an abductive account of Hodgkin and Huxley's reasoning.[14]

The overarching methodological point of this book is that if historians and philosophers of science are interested in how scientists confirm compositional hypotheses, then one important resource is the examination of the primary experimental literature. If they take this science-first approach – one that involves asking new questions – they will find that existing answers to other questions are unsatisfactory. Existing views need to be reconsidered in the light of new questions.

0.3 Roadmap

The book is divided into three parts. The first part presents the theory of singular compositional abduction and the portions of scientific reasoning that it is meant to describe. Chapter 1 will articulate and defend the

[11] See, for example, Craver (2007, 2008), Weber (2008), Craver and Darden (2013), Levy (2014), and Bickle (2023).

[12] Another important part of the story is the variation in the external sodium concentration. This will be discussed in Chapter 3.

[13] See, for example, the discussions in Craver (2007), Craver and Darden (2013), Baumgartner and Gebharter (2016), Baumgartner and Casini (2017), Craver et al. (2021), and Prychitko (2021).

[14] Baumgartner and Casini (2017) have an account of "constitutive abduction," but they do not discuss any scientific cases. Further, their model assigns no role for intralevel experiments. Still further, their account provides an abductive interpretation of the results of multiple scientific experiments, whereas mine provides an abductive interpretation of a single experiment.

Table 0.1 *Entities, their composition, and their use in explanation*

Type of entity	Compositional relation	Compositional explanation
Individuals	Lower level individuals together *constitute* a higher level individual	Singular Analytic Compositional Explanation
Activity instance	Lower level activity instances together *implement* a higher level activity instance	Singular Dynamic Compositional Explanation
Property instance	Lower level property instance together *realize* a higher level property instance	Singular Standing Compositional Explanation

metatheoretic presuppositions of this work, most notably the commitment to developing a theory that provides a descriptively adequate account of some of the reasoning scientists offer in the recent primary experimental literature in support of compositional hypotheses. Having a descriptively adequate theory is more than merely describing or reporting what scientists themselves say they are doing. It is more than describing or characterizing reasoning in laboratory life or reasoning that appears in review articles or textbooks. The primary experimental literature has a distinctive role in science and it is the focus of attention here.

Chapter 2 will present the principal features of a theory of singular compositional explanation. One could, of course, write an entire book on singular compositional explanation; however, this is not such a book. So, there will be many issues that do not receive the attention they would deserve in a book entirely devoted to scientific explanation. In brief, the theory is this. Singular compositional explanations are representations of natural ontological dependence relations among spatiotemporally localizable explanandum and explanans entities. Further, there is a plurality of singular compositional explanations corresponding to a plurality of singular compositional relations among explanandum and explanans entities. (See Table 0.1 for a summary.) There are the relatively familiar explanations of instances of activities of wholes in terms of instances of activities of the associated parts of the whole. These are "singular dynamic compositional explanations." Such explanations implicate a natural ontological dependence relation of "implementation." In addition, there are explanations of instances of properties of wholes in terms of instances of properties of their associated parts. These are "singular standing compositional explanations." Finally, there are explanations of individual wholes in terms of their individual associated parts. These are "singular analytic compositional explanations."

Whereas Chapter 2 outlines a relatively coherent body of theory regarding singular compositional explanation, Chapter 3 draws attention to a more motley collection of features of singular compositional explanations as found in the scientific experimental literature. So, for one thing, scientists sometimes explain the rate of an activity of a whole. As Hodgkin and Bernard Katz comment, "Our hypothesis therefore suggests that the rate of rise of the action potential should be determined by the rate of entry of sodium" (Hodgkin & Katz, 1949a, p. 55).[15] Second, scientists sometimes explain scientific results. They are not "mysterians" about how results come about. The theory of singular compositional explanation is part of a story about explaining scientific results. Finally, scientists often try to explain the results of controlled experiments. With closer attention to Hodgkin and Huxley's experiment introduced above, one sees that they are trying to explain why axon no. 15 *in seawater* displays an initial *inward* current when depolarized by 65 mV, whereas *in a sodium-free medium* it displays an initial *outward* current when depolarized by 65 mV. The "control" assumption is that there is only one difference between the conditions in which there is an inward current and the conditions in which there is an outward current. Further, however we might explain axon no. 15's behavior in the two contexts, the explanation is not simply a singular compositional explanation.

Chapter 4 will introduce the theory of singular compositional abduction built around the work on singular compositional explanation in Chapters 2 and 3. It begins by identifying features that singular compositional abduction shares with many other abductive inferences. It also provides an account of why some scientists would take abductive reasoning to be truth-conducive. Further, it explains how abductive confirmation offers a successor to the logical positivist theory of hypothetico-deductive (HD) confirmation. Hypothetically, if sodium ions were to move across the axonal membrane after a 65 mV depolarization, then that would make for an experimentally detectable initial inward current.[16] Given the experimentally detected current, scientists take themselves to have a good reason

[15] In one autobiographical account, Hodgkin (1994, p. 274) comments, "Another strong argument for the sodium theory was that the maximum rate of rise of the action potential was approximately proportional to the sodium concentration."

[16] In the abduction/IBE literature, it is common to suppose that possible explanations lead to actual explanations. See, for example, Lipton (2003), McCain (2015, 2019), Fumerton (2017), and Lange (2022).

to think that there are sodium ions moving inward across the axonal membrane soon after a 65 mV depolarization. For much of the twentieth century, empiricist philosophers would have interpreted this scientific practice as illustrating HD confirmation. I propose that it is an instance of singular compositional abductive confirmation. The broader conclusion is that some, but not all, scientific hypothetical reasoning that philosophers of science would have once taken to be HD confirmation is in fact singular compositional abductive confirmation.

The second part of the book, Chapters 5 and 6, will provide detailed case studies showing how the theory from the first part provides a descriptively adequate account of some of the scientific reasoning in some of the twentieth- and twenty-first-century scientific experimental literature. Chapter 5 will describe more of the scientific reasoning involved in developing the Hodgkin–Huxley theory of the action potential. Chapter 6 will review the ongoing debate regarding the neurophysiological basis of the Hermann grid illusion.

Although compositional abductive reasoning plays a part in both historical episodes, there are variations in how experimental work figures into that reasoning. Hodgkin and Huxley's research focused on the manipulation of axons and the subsequent measure of axonal voltages or currents. The results of these experiments were, often in combination with further assumptions and mathematical reasoning, used to develop compositional hypotheses about ion fluxes and membrane permeabilities. Sometimes, these inferences led Hodgkin and Huxley to hypotheses they found convincing. At other times, they did not.

Research on the Hermann grid has relied extensively on psychophysical methods wherein vision scientists present participants with one or another physical stimulus – some version of a Hermann grid – and then measure the participant's response. These intralevel results are generally combined with independent neurophysiological experiments that involve stimulating photoreceptor cells, sometimes one at a time, and then measuring downstream cellular responses. Both psychophysicists and neurophysiologists have their own interpretations, but my interest lies in the compositional interpretation of psychophysical results in terms of neurophysiological individuals, activity instances, and property/relation instances.[17] This is where scientists often rely on compositional abduction.

The first two parts of the book form the "primary project" of articulating and defending a positive account of several important instances of

[17] Hereafter I will typically simplify "property/relation instances" to "property instances."

scientific reasoning. The third part carries out a "secondary project" of describing possible rival accounts, how the rival accounts differ from my account, and some reasons for taking my positive account seriously. It probably cannot be emphasized too strongly that my project is *not* a matter of taking "off the shelf" ideas, such as Peirce's conception of abduction, Harman's concept of IBE, or some version of Carl Craver's manipulationism, and showing how scientific work fits into those preconceptions. My science-first approach looks closely at actual scientific reasoning, and then develops a theory that yields a descriptively adequate account of that reasoning. It should not be surprising that my version of a science-first approach leads to distinct ideas about confirming compositional hypotheses.

Chapter 7 will review how my account differs from that offered by Peirce and various "Neo-Peirceans." The key difference is that, whereas Peirce and the Neo-Peirceans assume that scientists use abduction to introduce new hypotheses, I propose that scientists use it to confirm hypotheses. Chapter 8 will illustrate respects in which the compositional abductive account differs from an application of the Harman picture of abduction to local scientific reasoning. Here I emphasize a distinction between abduction and warranted abduction, so that one can have abductive inferences that do not rule out rival hypotheses. The distinction facilitates the recognition that the results of individual scientific experiments, such as Hodgkin and Huxley's, do not typically warrant the adoption of some hypothesis. Instead, the results of a single experiment are only part of a more extended process of establishing a hypothesis. Chapter 9 will contrast the compositional abductive account of the scientific interpretation of controlled experiments with the account in Lipton (2003). The most significant difference between the views surveyed in Chapters 7–9 and the current view is the development of a specific theory of explanation and the use of extended case studies.

Chapter 10 will contrast the abductive approach with Carl Craver, Stuart Glennan, and Mark Povich's "matched interlevel experiments" (MIE) approach.[18] MIE is the latest iteration of a manipulability account of the testing of compositional hypotheses.[19] One of the most significant differences between the compositional abductive approach and MIE is that the former recognizes the use of intralevel experiments to confirm compositional hypotheses, whereas the latter does not. Chapter 10 also broaches an abductive approach to the interpretation of the results of some

[18] Craver et al. (2021).

[19] See, for example, Craver (2007), Kaplan (2012), Craver and Darden (2013), and Prychitko (2021).

interlevel experiments, wherein something at one level is manipulated and then something at another level is measured. One descriptive advantage of the abductive approach is that it recognizes that scientists sometimes confirm compositional hypotheses using a combination of intralevel and interlevel experiments.

Chapter 11 will summarize some of the principal conclusions of the book but also draw attention to a limitation of my account. In reading carefully through the primary experimental literature, one quickly finds that scientific practice is much more complicated than Peircean abduction, IBE, and manipulationism may have led one to believe. The practice of explaining rates of activity instances is an illustration. Further, once one recognizes a scientific practice, the historian and philosopher of science will want a theory of that practice. Yet, the recognition of practices is much easier than the development of a theory of those practices. In Chapter 4, I propose explanations of some rates, but leave open the possibility that there are other explanations of other rates. In this example, we see that the science-first methodology adopted here readily paves the way for philosophical theory revision. A more serious example involves functional magnetic resonance imaging (fMRI). The blood-oxygen-level-dependent (BOLD) signal in fMRI indicates changes in amounts of deoxyhemoglobin. Yet, the deoxygenation of hemoglobin is not typically taken to explain any cognitive processes. Deoxygenation is, instead, somehow "coupled" to the neurophysiological processes that implement cognitive processes. As familiar as this practice is, neither the abductive approach nor the MIE approach as they stand provides a theory of this practice.

CHAPTER 1

The Metatheory of the Project

First and foremost, this work is a bit of history and philosophy of science. It uses historical context and philosophical concepts, such as constitution, implementation, and abduction, to develop a theory of the scientific interpretation of experimental results in the recent primary experimental literature. It attempts to make explicit what is often merely implicit in published scientific research.

On the surface, this is a thoroughly pedestrian project, but its methodological starting point leads my theory to depart from many common expectations. There are so many departures, in fact, that the methodological stage setting in this chapter makes for one of the longest chapters of the book.

As described in Section 1.1, the theory introduces philosophical vocabulary and concepts, unconstrained by vocabulary and concepts familiar to scientists. One point of setting aside such a constraint is that the goal is not merely to report what scientists claim to be doing or think they are doing in one particular scientific domain. To get beyond describing scientific practice and on to a theory of the practices, it is often helpful to have new concepts and terminology.

Section 1.2 describes the exclusively descriptive character of the theory. Many projects in the philosophy of science propose to draw metaphysical and epistemological conclusions from scientific practice. Stated simplistically, a metaphysical argument might observe that scientists postulate X's, and from there infer that there are X's. Stated simplistically, an epistemological argument might observe that scientists use method Y to draw conclusions about the world, and from there conclude that method Y is justified and justificatory. Philosophers, of course, hedge these inferences, but I propose not to take either the ontological or the epistemological steps, even with hedges added. A principal part of the rationale for this is that the descriptive work is challenging enough.

Section 1.3 describes the theory's focus on the scientific interpretation of experimental results presented in the recent experimental literature. This is important, since a theory that focuses on what is happening at the moment of scientific insight, when a scientist first conceives an answer to a scientific question, would likely be quite different. A philosopher's theory of the scientific genesis of "happy thoughts" should probably draw on results from cognitive science. By contrast, the theory developed here will rely upon a close, granular reading of arguments in the recent scientific experimental literature. This work also backgrounds what is compiled into review articles or textbooks. It makes sense to bracket the reasoning in review articles since these often include reasoning that is not limited to that about experimental results. Further, textbooks, for their part, focus on communicating scientific conclusions rather than supporting scientific conclusions.

Section 1.4 describes the explanatory and abductive pluralism adopted here. Scientists provide many types of explanations leading them, in principle, to provide many types of abductive inferences. This section foreshadows the plurality of explanations and abductions discussed more concretely in Chapters 2–6.

Section 1.5 attempts to address some of the challenges facing the philosophical use of case studies, such as concerns about "cherry-picking" and overgeneralization. These are serious concerns that cannot be met simply by beginning with actual cases. To address the issue of generalization, the proposal developed here is that to generalize reasonably from a handful of case studies one should identify specific features of the historical cases that can guide generalizations.

1.1 Invoking Philosophical Concepts

By invoking "philosophical concepts" and "philosophical terminology," I reject a common assumption about how historians and philosophers of science should proceed, namely, that they should limit themselves to the concepts or vocabulary that scientists use. For my part, this is overly restrictive. Although I am sensitive to scientific terminology and will often draw attention to it, there are times when a departure from that terminology is in order. My goal is to develop a theory that accurately describes some of the scientific abductive reasoning in the experimental literature. Sometimes new terminology helps.

Sometimes, scientific terminology is well-suited to describing actual scientific reasoning. So, for example, scientists often use the terms "data" and "results" to describe at least a subset of the products of their experimental work. Further, in the case studies I examine, scientists use "account for" and "explain" interchangeably. I generally use "explain," because it is more commonly used by philosophers. So, in some instances, I largely embrace scientific terminology – fleshing out some of the scientific presuppositions – because the scientific terminology does the theoretical work I want to do.

As a point of clarification, I try to thread a needle in using the words "explain" and "explanation." On the one hand, I do *not* intend for the words "explain" and "explanation" to capture an English language, or even scientific language, use of the words "explain" and "explanation."[1] Thus, I do not mean to do lexicography. Nor, on the other hand, do I propose merely to stipulate what I mean by "explain" or "explanation." To repeat, my goal is to develop a theory of some of the abductive reasoning scientists use in connecting experimental results to compositional hypotheses regarding the world. This requires a theory of what I will call "explanation." My theory embraces certain ideas that philosophers have recently had regarding explanation, so it is not entirely detached from what has been popularly, scientifically, or philosophically meant by "explanation." Then again, it may have some features that are not found in some theories of explanation in the philosophical literature.

Although there are times when scientific terminology is well-suited to my project, there are also, unfortunately, times when it is not. This happens for very many reasons. Sometimes scientists do not adopt a consistent vocabulary, even within the space of a single paragraph or two. Günter Baumgartner wrote of *explaining* perception when viewing white bars on a black background but *understanding* perception when viewing black bars on a white background.[2] One philosophical interpretation of this is that Baumgartner thinks there is some important difference between the two cases, whereas the interpretation I adopt is that the terminological shift is merely stylistic. Semmelweis wrote of atmospheric-cosmic-terrestrial influences/conditions causing puerperal fever, operating on individuals, and occasioning an epidemic.[3] One might assume that Semmelweis attaches some significance to the difference

[1] For a sample of philosophers who are concerned with ordinary language use of "explain," and its cognates, see Jenkins (2008), Craver (2014), Şerban (2017), Wright and Van Eck (2018), and Craver and Kaplan (2020).

[2] Baumgartner (1960, p. 22).

[3] Semmelweis (1983, pp. 65–66).

between causing, operating, and occasioning. Alternatively, one might think, as I do, that the differences are merely stylistic.

Sometimes scientists have not developed a fixed, standard terminology that is useful for philosophical purposes. Scientists never use "explanans" or "explanandum" in the cases I examine, but those terms are extremely useful in developing a philosophical theory of explanation. In the cases I study, scientists never use "constitution," "realization," "implementation," and most notably "abduction," but I think that those terms/concepts are essential for understanding certain features of scientific reasoning. Just as one can speak in prose, reason using *modus ponens*, or commit a microaggression without having articulated the concepts of prose, *modus ponens*, or a microaggression, so scientists can invoke implementation or reason abductively without having those concepts.

There are also more subtle reasons for departing from extant scientific terminology. Sometimes there is a scientific term, but that term does not capture certain features that I take to be important. Here, I have in mind scientists writing about the interpretation of results. In general, "interpretation" is a term of ordinary and scientific language that can sometimes include the singular compositional explanation of results. As I read science, scientists sometimes interpret results using statistical methods, but sometimes they interpret them using singular compositional abduction. Thus, the interpretation of results covers more cases than does the singular compositional explanation of results. So, while I take it to be fine to discuss the interpretation of results, I also need more specific terminology that captures the additional features of the cases that I have in mind.

Another reason that some scientific terminology is sometimes not suited to my purposes is that it is used only within one area of research. I want to develop a theory that covers multiple distinct areas of research. Hodgkin and Huxley proposed that sodium ions carried the initial inward current of the action potential. This term "carried" describes a relation between an activity instance of an axon and the collective activity instances of multiple sodium ions. While there continue to be uses of this term for this purpose, it has not been exported to other areas of scientific research.[4] When Günter Baumgartner proposed that retinal ganglion cells (RGCs) provide part of the biological basis of the Hermann grid illusion, he did not say that retinal ganglion cells carry the Hermann grid illusion. Insofar as one wants a single term to describe both the Hodgkin–Huxley case and the Baumgartner case, one will have to improvise. My improvisation is to write

[4] Hammond (2014, p. 35) mentioned ions carrying currents.

of implementation. This term covers a relation I take to be found in multiple areas of scientific investigation.

Perhaps the most problematic cases where scientific terminology does not suit my purposes is when scientists in my case studies use "cause" or "effect," where I would describe what they are doing in terms of composers and composed. So, for example, Hodgkin and Huxley claim that "the rising phase of the action potential is caused by the entry of these ions" (Hodgkin & Huxley, 1952a, p. 449). A common presupposition is that if scientists call something a "cause" or an "effect," then these are the terms that philosophers should use. This is a defensible view, so for that reason I spend a fair amount of time supporting an alternative. Chapter 2 has a substantial discussion that motivates drawing a distinction between causal relations and compositional relations. Here, I suppose that this is simply another case where scientists do not make a distinction that is important for accurately describing what they are doing.

For what it is worth, at least some New Mechanists also implicitly reject the "stick to the scientific language" restriction. Many New Mechanists focus on something like characterizing the extension of the scientific use of the term "mechanism," so in this case try to abide by scientific terminology. Nevertheless, some New Mechanists rely on concepts and terminology, such as mechanistic constitution, constitutive relevance, mutual manipulability, and constraining a hypothesis space, that are not native to scientific discussion.[5] These are not naturally occurring concepts/terminology in scientific discussion. This is not to say that there are no scientists who have used these philosophical concepts or terminology. Some scientists have read the New Mechanist literature and embraced parts of it.[6]

Perhaps what motivates the restriction of philosophical concepts and terminology to scientific concepts and terminology is the sense that going beyond the scientific concepts and terminology amounts to an unwarranted philosophical interpretation. Here, my counterproposal is that there are interpretations that are warranted and interpretations that are not warranted. Indeed, at many points in my discussion, I will present objections to other philosophical interpretations of scientific reasoning. My objections will not, however, be that these interpretations are interpretations. Instead, the objections will be that the rejected interpretations somehow do not correctly describe the cases they may be thought to describe. The entire crux of determining which interpretations are

[5] Craver (2007), Craver and Darden (2013).

[6] Japyassú and Laland (2017) is an instance.

warranted and which are not is an *a posteriori* issue that has to be settled on a case-by-case basis.

My method is not merely to report what scientists write in the experimental literature; I offer a theory of why they write what they do. They sometimes write as they do because they are making use of singular compositional abduction. Here, an analogy might be helpful. Sociologists and social psychologists sometimes distinguish between ingroups and outgroups. An ingroup is roughly a group with which an individual psychologically identifies, such as a race, religion, or culture. In contrast, an outgroup is a group one does not psychologically identify with. Members of the groups may not be aware of the ingroup–outgroup distinction. Still, sociologists might use the distinction to explain the observation that a comment in the mouth of a member of one's ingroup provokes a different reaction in the mouth of a nonmember. Other sociological concepts, such as "implicit bias" and "racial resentment," might not be familiar in a given human population but might nevertheless be scientific concepts that provide illuminating explanations of various sociological phenomena in that population.

Like the sociologist or social psychologist, a historian and philosopher of science might develop concepts such as realization, implementation, constitution, singular compositional explanation, singular compositional abduction, etc., to illuminate the work of scientists. Moreover, just as individuals in a society may not have articulated concepts of ingroup and outgroup, so the community of scientists may not have articulated concepts of realization, implementation, and so forth. The goal of this work is in this spirit. It attempts to render explicit the commitments often implicit in actual scientific theorizing and argumentation.

The analogy here is perhaps not perfect. Scientists conduct research, and are trained to conduct research, on entities in the natural world, such as action potentials. In addition, in their research and training, they develop a sense of what they are trying to do and how they are trying to do it. Roughly speaking, they have views about the natural world as well as views about science, its methods, and its products. They have views about science and, perhaps, a philosophy of science. Scientists may, therefore, be more self-conscious about what they do in their scientific work than average individuals are about what they do in their social interactions.

A potentially more worrisome observation is that scientists might have a philosophy of science that conflicts with what a historian and philosopher of science might have to offer. Anecdotally speaking, Popperian falsificationism is quite popular among scientists.[7] For another example, Hodgkin and Huxley once claimed that "the rising phase of the action potential is caused by the entry of these [sodium] ions" (Hodgkin & Huxley, 1952a, p. 449). However, on the theory proposed here, the relation between the rising phase of the action potential and the entry of sodium ions is not causation but implementation. The justification for my proposal is that there are features of the implementation relation that differ from features of the causation relation. Most notably, the rising phase of the action potential is contemporaneous with the influx of sodium ions into the axon.

In the face of such conflicts, I do not assume that the historian and philosopher of science must be mistaken. The scientist could be mistaken. I do not suppose that this is because philosophers have some special insight into the nature of reality. Nor do I suppose that philosophy or metaphysics somehow "sets the limits of scientific theorizing."[8] Instead, the philosopher of science could get it right about how science works simply by devoting more effort to figuring it out. Where scientists might focus their attention on the rising phase of the action potential, historians and philosophers of science might devote their attention to what is implicit in the relation between ion fluxes and action potentials. Such detailed consideration has led many philosophers to distinguish causal relations from compositional relations, even though scientists may sometimes say that lower level activities cause higher level activities. In the end, the best that one can hope for, perhaps, is some measure of agreement between what scientists say about science, its methods, and its products, and what historians and philosophers of science have to say. The possibility of such agreement, at least, is presupposed in this book.

1.2 An Exclusively Descriptive Account of Scientific Reasoning

My project is to provide a descriptively adequate account of published scientific reasoning. In selecting this project, I intentionally set aside metaphysics and epistemology. I will not make claims about the fundamental nature of reality or what scientists may, or may not, justifiably believe true. Instead, I offer an account of what I take to be some scientific commitments and practices. The simple and primary reason for this is that

[7] Raman and Ferster (2022, p. 69) interpret Hodgkin and Huxley (1952a) as engaging in falsificationism at one point.

[8] Heil (2012, p. 11).

the descriptive project is challenging enough without trying to support metaphysical and epistemological conclusions.

1.2.1 Some Metaphysical Presuppositions of Science

My account assumes that scientists are ontologically committed to, among other things, the existence of individuals, property instances, activity instances, and ontological dependence relations among them. The account proposes that certain ontological, hence metaphysical, commitments or presuppositions are found in science. The account does not, however, offer ontological or metaphysical commitments *simpliciter*. For, it is one thing to say that a scientist thinks that X exists and quite another to say that X exists. Metaphysics, as I understand it, concerns the latter existence claims.

One can, of course, forge a link between scientific metaphysical commitments and metaphysics. Tim Maudlin offers the following argument:

> Metaphysics is ontology. Ontology is the most generic study of what exists. Evidence for what exists, at least in the physical world, is provided solely by empirical research. Hence the proper object of most metaphysics is the careful analysis of our best scientific theories (and especially of fundamental physical theories) with the goal of determining what they imply about the constitution of the physical world (Maudlin, 2007, p. 104).

I happen to share Maudlin's view that our best-supported metaphysics is the metaphysics in science, so that philosophers should accept the ontology found in science. This, however, is not a claim I support in this book. The careful analysis of scientific theories Maudlin alludes to is challenging enough, so I limit myself to that.

As a note of clarification, I might add that Stavros Ioannidis and Stathos Psillos agree with the idea that one can describe scientific metaphysical commitments without endorsing them as *bona fide* metaphysical commitments.[9] They may be alone among New Mechanists in adopting this view. Nevertheless, they go too far when they deny that there are arguments such as Maudlin's: "Concerning the question of metaphysics, one can, surely, give reasons to prefer some particular metaphysical picture over others. Our point is that whatever else these reasons are, they are not related to scientific practice" (Ioannidis & Psillos, 2022, p. 6). Whatever objections one might have to Maudlin's argument, one should not include the claim that it is unrelated to scientific practice. At the very least, I part ways with Ioannidis and Psillos on this.

[9] Ioannidis and Psillos (2022, p. 6).

Although many philosophers are inclined to dismiss an exclusively descriptive metaphysical project, this inclination underestimates its difficulty. An accurate description of scientific ontological commitments is not a trivial matter. Let me provide an illustrative example concerning the scientific commitment to property instances, something about which New Mechanists have had relatively little to say. My official ontology for singular compositional explanation involves property instances, rather than properties. I will justify this in Chapter 3, but for the next few paragraphs I will run the discussion in terms of properties. This will simplify matters. Moreover, the point I make regarding properties will carry over to property instances.

Sometimes the "New Mechanism," as opposed to one or another form of "old mechanism," is traced to Stuart Kauffman's idea that "Typical explanations in biology exhibit the manner in which parts and processes articulate together to cause the system to do some particular thing" (Kauffman, 1970, p. 257).[10] New Mechanists have, accordingly, for decades, largely limited themselves to two ontological categories corresponding to Kauffman's "parts" and "processes." Properties, however, are mentioned only on the periphery.[11]

The discussion in Machamer et al. (2000) is illustrative. They famously claim that "Mechanisms are entities and activities organized such that they are productive of regular changes from start or set-up to finish or termination conditions" (Machamer et al., 2000, p. 3). There is no mention of properties here. Quickly thereafter, however, they propose that "Mechanisms are composed of both entities (with their properties) and activities. Activities are the producers of change. Entities are the things that engage in activities. Activities usually require that entities have specific types of properties" (Machamer et al., 2000). Glennan and Illari (2017) have more recently illustrated this same pattern. They propose that "A mechanism for a phenomenon consists of entities (or parts) whose activities and interactions are organized so as to be responsible for the phenomenon" (Glennan & Illari, 2017, p. 1), but also quickly thereafter note that "Metaphysical mechanists ... likely see mechanisms as central to understanding some traditional metaphysical questions, e.g., about the nature of

[10] For a brief history, see Wimsatt (2018).

[11] One notable exception is Piccinini (2020), which often refers to properties. Piccinini's approach, however, is apparently driven by an attempt to integrate the philosophical literature on realization (which often focuses on properties) and the literature on New Mechanism (which largely ignores properties). Piccinini's approach is not driven by fidelity to scientific practice. He does not argue that a descriptively adequate account of scientific practice should recognize a role for properties.

objects and properties, parts and wholes, causal relations and laws of nature" (Glennan & Illari, 2017, p. 2).

What scientific practice, one might ask, indicates a role for properties? Craver and Darden (2013) rightly mention that properties enable activities. So, they observe that "[e]ntities have properties, such as structure and orientation, that enable them to engage in specific activities" (Craver & Darden, 2013, p. 16).[12] Suppose that properties do enable activities. *Prima facie* there should be a vast number of property-invoking explanations of why certain entities engage in certain activities. There should be a vast number of "enabling explanations." So, one might ask, "Why does water expand upon freezing?" The answer has to do with the polarity of water molecules. Why does the movement of sodium ions into the axon generate a current? The answer has to do with sodium ions having positive charge. There is, however, more to be said than Craver and Darden mention. Given that water molecules are polar, one should expect there to be a scientific explanation of their polarity. Why, one might ask, are water molecules polar? Although New Mechanists have rarely considered such cases, there is a compositional explanation of this property of water molecules in terms of the properties of its parts. The explanation of the polarity of a water molecule is based on the relative electron affinities of hydrogen and oxygen. An oxygen atom is more electronegative than a hydrogen atom – electrons tend to cluster on the oxygen side of the molecule – thereby making the oxygen side of the water molecule more negative. This line of thought suggests that (a) for every explanation of an activity of a composite individual, there will be a property that enables that activity and (b) that there will be a scientific compositional explanation of the activity-enabling property of the composite individual. Maybe this is mistaken, but it is a line of thought that merits some philosophical consideration. I will return to enabling in Chapter 3.

To summarize this subsection, this book offers a purely descriptive theory of what scientists take to be involved in the singular compositional abductive interpretation of scientific results in the experimental literature. This departs from a common move in the philosophy of science to infer from scientific commitments regarding the way the world is to metaphysical conclusions regarding the way the world is. In this book, I do not take this further step, since I think the accurate description of actual scientific commitments is a worthy project in its own right. As an illustration of this challenge, I have drawn attention to the decades-long New Mechanist

[12] See, also, Craver and Darden (2013, pp. 17–18, 78–79, 152).

practice of relative inattention to properties. On my theory, individuals, property instances, and activity instances are "equal ontological players" in scientific reasoning. They are ontologically intertwined in scientific thinking. New Mechanists have not accurately described this "ontological egalitarianism."

1.2.2 *Some Methodological Presuppositions of Science*

As my focus is on describing scientific compositional abduction as found in the primary experimental literature, I will set aside any number of issues that have occupied epistemologists and philosophers of science for decades. Thus, I do not consider whether abduction or IBE is a "fundamental rule of inference."[13] I will not be concerned with whether warranted enumerative induction is a special case of warranted abduction.[14] I do not consider whether any specific abductive inference is justified or whether abduction, in general, is justified.[15] One topic I will completely avoid is Bayesian confirmation. Although I find Bayesianism implausible as an account of the scientific reasoning I examine, I find that the pros and cons of Bayesianism are so well-worked that I do not see that I have anything to add.[16]

Focusing exclusively on the descriptive project sets this book apart from many other works in the philosophy of science that propose to use actual scientific reasoning as a guide to how scientific justification proceeds. New Mechanists often combine the study of the descriptive and the normative, but they are not alone in this. Lindley Darden has proposed that the history of science might serve as a source of "compiled hindsight."[17] William Bechtel and Robert Richardson propose to look at actual scientific reasoning, seeing how it succeeds and fails, for a guide to how scientists should reason.[18] Craver provides a lengthy defense of providing both a descriptive and normative account of mechanistic explanations in neuroscience.[19] Among logicians of abduction, Dov Gabbay and John Woods have proposed an "*Actually happens rule.* To see what agents should do, look first to what they actually do. Then repair the account if there is

[13] Cf., for example, Harman (1965, 1968), Ennis (1968), Fumerton (1980, 2017), Lipton (2003), Mackonis (2013), Khalifa et al. (2017), and Williamson (2016).

[14] Cf., for example, Harman (1965, 1968), Ennis (1968), and Fumerton (1980).

[15] For a small sample of this literature, see Van Fraassen (1980), Psillos (1996), Ladyman et al. (1997), Okasha (2000), Weber (2004), Carter and Pritchard (2017), and Douven (2022).

[16] For a very tiny sample, see Glymour (1980a), Earman (1992), Howson and Urbach (2006), and Norton (2021).

[17] Darden (1987, p. 39), Cf., Craver (2007, p. x) and Craver and Darden (2013, p. 10).

[18] Bechtel and Richardson (2010, pp. 10–11).

[19] Craver (2007, pp. vii–xi).

reason to do so" (Gabbay & Woods, 2005, p. 5). John Norton also looks closely at actual science and uses that as a basis for conclusions about warrant or justification.[20]

One reason not to pursue the descriptive and normative projects simultaneously is that it is a lot of work to put into practice. Although Gabbay and Woods advocate the *actually happens rule*, they do not discuss any empirical evidence about how agents actually reason. To flesh this out a bit with an example, Gabbay and Woods follow Peirce in assuming that abduction is "ignorance preserving." They do not, however, cite any empirical work in support of the view that scientists, or any others, treat abduction as ignorance preserving. I have proposed that Hodgkin and Huxley reasoned abductively about the sodium hypothesis. After they performed all those experiments, did they not know that sodium carries the initial inward current of the action potential? There are times when they write about what they have supported or established regarding how the action potential works.[21] Nevertheless, this is not a question Gabbay and Woods try to address.

A second reason not to pursue the descriptive and normative projects simultaneously is that matters quickly become complicated. Peirce proposes that abduction provides no confirmation. Suppose, as will be argued in Chapter 5, that Hodgkin and Huxley understood abductive reasoning to confirm the hypothesis that the initial inward current was generated by sodium. In other words, suppose that Hodgkin and Huxley assumed that abduction provides confirmation. Then there is a *prima facie* conflict. Peirce apparently had one assessment of abductive reasoning; Hodgkin and Huxley apparently had another. What is to be made of it? There are many possibilities. Perhaps Peirce made a descriptive mistake about how scientists reason. Or, maybe he did not mean to describe how scientists reason, but how scientists should reason. Maybe then Hodgkin and Huxley were drawing an unwarranted conclusion. But, if one concludes that Hodgkin and Huxley were in some sense wrong to draw this conclusion, then why were they wrong? What basis is there for saying that Hodgkin and Huxley were making a mistake? And, what if one finds that many scientists break with a normative interpretation of Peirce's proposal, as by awarding Hodgkin and Huxley the Nobel Prize for their work? Does this mean that there is a lot more scientific "malpractice" than one might have anticipated?

[20] Norton (2021).

[21] See, for example, Hodgkin and Huxley (1952a, p. 471) and Hodgkin and Huxley (1952d, p. 541).

The point here is not that these questions cannot be resolved. It is that they involve complexities that I propose to bracket in this book. Indeed, they are questions that have typically been set aside in the literature. There are numerous discussions of Peirce's idea of abduction as merely generating hypotheses, but many do not consider what to say should descriptive-normative conflicts arise.[22] And Peirce is not the only one to have advanced apparently descriptive and/or normative claims. Indeed, part of what enables many philosophers of science to think they can simultaneously address both descriptive and normative issues is that they do not examine cases in which proposed normative claims appear to depart from scientific practices. Matters are a lot simpler if one does not entertain descriptive-normative differences.

A third reason to begin with an exclusively descriptive account of scientific reasoning is that some philosophers working on abduction are inclined to begin with the normative. Many philosophers begin with Harman's picture of warranted abduction and picking a best explanation on the basis of simplicity, plausibility, etc. Inspired by this picture, they might assume that, in using their first experimental results, Hodgkin and Huxley ruled out rival hypotheses by appealing to sundry explanatory virtues such as simplicity, plausibility, etc. If one reads through the text of Hodgkin and Huxley's paper discussing the sodium hypothesis experiments, however, one sees that Hodgkin and Huxley consider no rivals and there is no explicit mention of simplicity or plausibility.[23] One might then propose that Hodgkin and Huxley are *implicitly* considering rival hypotheses and *implicitly* ruling them out by appeal to simplicity and plausibility. But what is the evidence that Hodgkin and Huxley implicitly rule out rivals? Is this just an *ad hoc* presupposition one makes in order to apply to Harman picture?

Why should a historian and philosopher of science take this path? Why begin with Harman's picture of warranted abduction – a picture developed for a project in epistemology – as a starting point for a project in the history and philosophy of science? Why begin with such a preconception? A casual familiarity with scientific research suggests that, at any given time, scientists believe that some rival hypotheses have been eliminated, but that

[22] See, for example, Hoffmann (1999), Niiniluoto (1999), Magnani (2001), Gabbay and Woods (2005), Schurz (2008), Campos (2011), Plutynski (2011), Mackonis (2013), McAuliffe (2015), Misak (2017), and Mohammadian (2021). None of these accounts explain why Peirce thought that abduction merely introduces hypotheses (worthy of pursuit). Nor do they consider whether Peirce was, in fact, correct in thinking this.

[23] More specifically, see the discussion at Hodgkin and Huxley (1952a, pp. 450–452).

others have not. So, there is some sense in which some rival hypotheses have implicitly been eliminated. Then again, a large part of actual science involves entertaining rival hypotheses and designing and performing experiments that will rule them out. In an important sense, many scientific inferences, such as those by Hodgkin and Huxley, may not be warranted to Harman's standards at the time they are made. Some rival hypotheses may have yet to be addressed. Instead, such warrant as scientists may have for some hypothesis may emerge, if it ever does, through a sustained diachronic process of scientific research. Indeed, Huxley himself had something like this thought. In autobiographical reflections some forty plus years after the publication of the Hodgkin–Huxley model, he wrote,

> Most of my life has been spent in posts where I was teaching medical students, and I am very glad to have had even that single year of clinical study since it enabled me to see how different the attitude of the clinician has to be from that of a scientist: the clinician has to make immediate decisions, often on slender evidence, while the scientist has (almost) unlimited time. (Huxley, 2004, p. 293)

While there is plenty of room to interpret what Huxley meant here, there is enough to suggest that historians and philosophers of science take seriously the idea that scientific justification often arises, when it does, not merely from the results of a single experiment.[24] Thus, rather than beginning with the Harman picture, I propose to look at actual published science without beginning with Harman's preconception. This is part and parcel of my science-first approach.

1.3 The Recent Primary Scientific Experimental Literature

Whereas this book is focused, first and foremost, on a descriptively adequate theory, it does not aspire to a descriptively adequate account of the whole of science. Nor does it aspire to a theory of the whole of scientific reasoning or even scientific inductive reasoning. To a first approximation, it focuses on a descriptively adequate account of how scientists sometimes justify compositional hypotheses in the recent primary scientific experimental literature. The qualifier "first approximation" is meant to cover a number of issues. For one thing, I do not have a precise

[24] Huxley's comment also invites a serious examination of whether abductive inference in medicine might be different than abductive inference in science. Maybe on limited evidence, the clinician must typically turn to simplicity, plausibility, etc., whereas the scientist typically has time to do another experiment.

characterization of "recent." Post-1930 is in the ballpark, but there is no specific event that suggests 1930, rather than 1925 or 1935 or 1945. Moreover, there are times when nineteenth-century science provides some helpful illustrations and helpful background. Finally, there are times when it is helpful to stretch the boundaries of "experimental literature" a bit.[25]

This book draws attention to abduction as scientists use it in the scientific literature. My engagement with science is, thus, what N. R. Hanson once derisively referred to as a "Logic of the Finished Research Report" (Hanson, 1960, p. 92). Of course, the engagement here is not so much with what I would call a "logic" *per se* but, more modestly, with a characterization of the justifications – the reasons – scientists sometimes provide for their hypotheses regarding what is compositionally related to what in the primary experimental literature. Adopting this focus, I, therefore, set aside abductive inference as it may be used in other domains, such as computer science, legal reasoning, and philosophical arguments for scientific realism.[26]

In focusing on finished research reports, this book focuses on a fragment of the reasoning found in "unsettled, dynamic, research sciences" (Hanson, 1958, p. 1). It focuses on some of the initial interpretations of experimental results for publication in the experimental literature. It, therefore, sets aside what is sometimes described as the "logic of discovery," understood as the psychology of individual scientists at the moment of scientific insight.[27] The present work is not concerned with the likes of what was passing through Archimedes's mind as he sat in his bathtub nor through Kekulé's mind as he sat by his fire. Nor is it concerned with the thinking that precedes those insightful moments. This book also sets aside the "packaging" of scientific theory and reasoning into review articles and textbooks. Nor is it focused on a somewhat more abstract concept of theory confirmation. The rationale for this focus merits articulation.

1.3.1 The Moment of Insight

Many philosophers will suppose that a philosophical theory of abduction will focus on scientists' initial conception of hypotheses. Peirce is widely credited with having distinguished abductive reasoning from inductive and

[25] Here I have in mind Hermann (1870), Baumgartner (1960), and Schiller and Carvey (2005).
[26] Magnani (2001), Psillos (2005).
[27] New Mechanists sometimes use the term "discovery" to cover both the initial advancement of a hypothesis and its subsequent confirmation. See, for example, Craver and Darden (2013). For discussion of this usage, see Schickore (2018).

deductive reasoning. He believed that abduction was the only means by which new ideas enter science. Once abduction has introduced a new hypothesis, deductive and inductive reasoning take over to confirm that hypothesis.[28] Hanson embraced this role for abduction, and Darden apparently carries it over into her brand of New Mechanism.[29]

Although linking abduction to the "moment of insight" will be natural to some, that is not a linkage I pursue. I take seriously the possibility that the process, or processes, underlying the initial psychological formulation of hypotheses might be distinctly different from the reasoning that goes into the interpretation of the results in the experimental literature. Since New Mechanists first proposed to study scientific discovery some decades ago, psychologists have developed an extensive body of research on human reasoning. As the result of experimental work on, for example, statistical and syllogistic reasoning, psychologists have offered multiple versions of what is often called a System 1/System 2 distinction. In one popularization of the idea, Daniel Kahneman briefly describes these:

> System 1 operates automatically and quickly, with little or no effort and no sense of voluntary control.
> System 2 allocates attention to the effortful mental activities that demand it, including complex computations. The operations of System 2 are often associated with the subjective experience of agency, choice, and concentration. (Kahneman, 2011, pp. 20–21)

System 1 operates quickly, as it apparently acts on various heuristic assumptions – biases – that violate principles of valid statistical and logical reasoning. One should, therefore, take seriously the possibility that the reasoning involved in the initial formulation of a scientific hypothesis might be performed by System 1, whereas the reasoning involved in the interpretation of the results in the experimental literature might be performed by System 2.

I do not here mean to explain and defend any version of this analysis; that is why I propose instead merely that the possibility needs to be taken seriously. Taking the proposal seriously would mean reviewing a large psychological and philosophical literature on human reasoning and explaining and defending one or a family of analyses. While undoubtedly an important philosophical project, it is not this project. It is a project that should rely quite heavily on psychology but less heavily on the close reading of scientific texts. It makes sense to separate these two projects.

[28] Peirce (1992). [29] Hanson (1960), Craver and Darden (2013, p. 81), Darden (2017, p. 255).

Here is another way of emphasizing the importance of the distinction between laboratory life and journal articles. In the IBE tradition, it is common to think of IBE in so general a fashion as to cover both what happens in the spontaneous scientific inferences in the laboratory and what is published more formally in scientific journal articles. Harman and Lipton apparently do this. In the history and philosophy of science, however, it is commonly held that what appears in scientific journal articles is not a faithful representation of what takes place in the lab. In a famously provocative paper, "Is the scientific paper a fraud?," Peter Medawar argued that scientific papers are not faithful representations of laboratory experimentation.[30] Hodgkin had his own version of this idea, touching on themes of chance and design in science:

> I believe that the record of published papers conveys an impression of directness and planning which does not at all coincide with the actual sequence of events. The stated object of a piece of research often agrees more closely with the reason for continuing or finishing the work than it does with the idea which led to the original experiments. In writing papers, authors are encouraged to be logical, and, even if they wished to admit that some experiment which turned out in a logical way was done for a perfectly dotty reason, they would not be encouraged to 'clutter-up' the literature with irrelevant personal reminiscences. (Hodgkin, 1976, p. 1)

I think there are likely to be differences between reasoning in the lab and reasoning in publications. We should not assume that they are the same. It might turn out that some general theory of abduction/IBE applies to both laboratory reasoning and experimental reasoning, but, for the space of this book, I will remain noncommittal about laboratory reasoning and focus exclusively on the experimental literature. In adopting this focus, I do not mean to weigh in on a debate about the relative importance of lab life versus publication. As I see it, laboratory experimentation and publication both have essential roles in science. It is, therefore, perfectly reasonable to focus on an account of the reasoning scientists offer in the experimental literature without trying to assess its importance relative to laboratory work or laboratory reasoning.

1.3.2 Scientific Reviews

What about later scientific work in reviews? In review articles, many important details of scientific reasoning are omitted for the sake of

[30] Medawar (1963).

characterizing the experimental and theoretical lay of the land regarding a particular topic at a particular time. To give an example, review articles will sometimes not mention experimental controls that are commonly employed in the primary experimental literature. In the many studies of rat spatial navigation through mazes – to provide a concrete example – steps are taken to ensure that the rats cannot rely upon olfactory cues from the surfaces of the maze to find their rewards. Such steps might involve wiping down the surfaces of the maze. As this is common knowledge and practice in the area, review articles will often not mention them. Nevertheless, the fact that scientists regularly use experimental controls is important for the abductive approach, since one rationale for an experimental control is that it enables one to rule out alternative explanations of experimental results. With appropriate controls in place, one cannot explain a rat's navigation abilities by appealing to olfactory cues, since olfactory cues would not have guided the rats to their rewards.

1.3.3 Textbook Science

Textbooks also often omit much of the reasoning found in the primary experimental literature, as their burden is not so much to convince students of the truth of scientific hypotheses as it is to transmit to them what the hypotheses are. To return to our example of Hodgkin and Huxley's 1952 experimental work, Hodgkin and Huxley performed three experiments that involved varying the concentration of extracellular sodium and the potential at which the axon was clamped. Textbook accounts of their work may not describe any of these experiments or the reasoning behind them on the assumption that this level of historical detail is unlikely to be needed by neuroscience students.[31]

As a second illustration, consider the textbook discussions of the action potential in Shepherd (1988) and Kandel et al. (2000), alongside some of the reasoning offered in support of the theory circa 1950. Around 1950, it was known that sodium ion concentrations were much higher outside of the cell than inside and that potassium ion concentrations were much higher inside of the cell than outside. Hodgkin and Huxley, among others, proposed that features of the action potential could be explained by postulating fluxes of these ions. As will be spelled out in more detail in

[31] Cf., for example, Shepherd (1988, chapter 6), Kandel et al. (2000, p. 151), and Hammond (2014, chapter 4). In recent years, however, neuroscientists have taken greater interest in these papers. See Raman and Ferster (2022) and Brown (2020).

Chapter 5, abductive methods were at work here. Hodgkin and Huxley's so-called indirect electrophysiological methods were complemented by "direct" methods, reported in Keynes (1948, 1949), Rothenberg (1950), and Keynes and Lewis (1951a, 1951b). In these latter papers, scientists argued that radioactively labeled sodium and potassium crossed the membrane during action potentials. For sodium, the experiments stimulated neurons in an external medium containing radioactively labeled sodium ions. When the cell was subsequently transferred to a sodium-free medium, it was found that the cell contained radioactivity that it must have carried from the external sodium ions. For potassium, the cells were loaded with radioactive potassium by keeping them in an external medium rich in radioactive potassium. After transferring the cells to a non-radioactive medium, then stimulating them, it was found that radioactive potassium was released into the medium. None of this experimental detail is reported in either Shepherd (1988) or Kandel et al. (2000). Hammond (2014) mentions the radioactive tracer work, but cites Hodgkin and Keynes (1955). Without examining the original research, the philosopher of science might overlook important instances of scientific reasoning. Chapter 5 will review a number of "minor" scientific inferences that do not make their way into textbooks, but that illustrate compositional abductive reasoning in action.

1.3.4 Theory Confirmation

I propose singular compositional abduction as part of an account of the scientific interpretation of certain experimental results. This focus distinguishes my account from one way in which accounts of abduction – generally understood as some form of IBE – may be related to scientific practice. Sometimes IBE is treated as an account of hypothesis confirmation, but it is also sometimes treated as an account of theory confirmation. Paul Thagard's work provides an early illustration, although it is framed in terms of theory choice. Thagard observes that,

> The phrase "inference to the best explanation" is relatively new, but the idea is old. Inference to scientific hypotheses on the basis of what they explain was discussed by such nineteenth-century thinkers as William Whewell and C. S. Peirce, and earlier still by David Hartley, Leibniz, and Descartes. To put it briefly, inference to the best explanation consists in accepting a hypothesis on the grounds that it provides a better explanation of the evidence than is provided by alternative hypotheses. We argue for a hypothesis or theory by arguing that it is the best explanation of the evidence. (Thagard, 1978, p. 77, cf., p. 76)

The discussion, up until the final sentence, is in terms of hypotheses. Thagard's subsequent discussion, however, ranges over Charles Darwin's theory of evolution by natural selection, Antoine Lavosier's oxygen theory of combustion, and Christian Huygen's wave theory of light. Further, in working out his account of theory choice, Thagard proposes that scientific theory choice is a function of, among other things, a theory's explanation of "facts."[32] This shift between hypotheses explaining and theories explaining is common in the literature.[33]

Singular compositional abduction is more plausible as an account of hypothesis confirmation than it is as an account of theory confirmation. Without venturing a characterization of the difference between hypothesis confirmation and theory confirmation, I appeal to the idea that theory confirmation is a function of – in a deliberately broad construal of the phrase – hypothesis confirmation. Theory confirmation, thus, to speak metaphorically, must cover more ground. In covering more ground, one is likely to find a much broader range of considerations than one finds in covering less ground.

To make this point more concrete and less metaphorical, consider the Hodgkin–Huxley–Katz papers circa 1950. These papers brought many experimental results to bear in support of the Hodgkin–Huxley theory of the action potential. In these papers, they mention simplicity from time to time. So, for example, Hodgkin and Katz comment that "A simpler type of hypothesis has recently been worked out, in collaboration with Mr Huxley, and forms the theoretical background of this paper" (Hodgkin & Katz, 1949a, p. 37).[34] Maybe there is some theory according to which the simplicity of a hypothesis counts in favor of the simplicity of an explanation.[35] And maybe this could be incorporated into a version of IBE. Thus, a complete account of how Hodgkin, Huxley, and Katz made their case for their theory of the action potential would have to include an account of the role they thought simplicity to have played. In contrast, if one examines, say, Hodgkin and Huxley's interpretation of the first

[32] Thagard (1978, p. 79f). For further, discussion of Thagard's account of theory choice, see Chapter 8.

[33] Okasha (2002, 22f), McCain (2019, 55f), Lange (2022, p. 85). The shift is more diffuse in Lipton (2003) and Bird (2022).

[34] See, also, Hodgkin and Katz (1949a, pp. 53, 55, 66), Hodgkin and Huxley (1952d, p. 505). Hodgkin, Huxley, and Katz also use "simple" in other ways that appear to have nothing to do with explanation. So, for example, there is this: "There is no simple way of estimating the error introduced by this approximation" (Hodgkin & Huxley, 1952d, p. 505).

[35] Thagard's theory of simplicity, however, does not appear to be such a theory, as it is based on the size and "nature" of the set of auxiliary hypotheses used in an explanation (Thagard, 1978, p. 86).

experiment of Hodgkin and Huxley (1952a), one will find no mention of simplicity. This provides *prima facie* reason to think that, while a singular compositional abductive account that does not involve simplicity will not have sufficient theoretical resources to account for the whole of the confirmation of the Hodgkin–Huxley theory of the action potential, it might be more at home in the narrow context of the interpretation of the results of Hodgkin and Huxley's first experiment.[36] These are themes to which I will return in Chapters 8 and 9.

1.3.5 The Primary Experimental Literature

In focusing on the argumentation in the primary experimental literature, I do not mean to deny an important role for moments of insight. What would science be without them? Nor do I deny the important role of review articles and textbooks. Each of these parts of the scientific enterprise has a role to play, but those roles are not the same. Putting matters quite roughly, publication of experimental results is the empirical input to the scientific community's theorizing; review articles attempt to package experimental and theoretical work into more intellectually manageable frameworks; textbooks are meant to communicate current scientific views to a new generation of scientists. Moreover, coming to understand these different roles may require different methods. As proposed below, a close reading of scientific texts – be they in the primary experimental literature, reviews, or textbooks – can plausibly give philosophers some insight into how scientists make their case for empirical hypotheses. However, such methods might not be adequate for understanding the psychology of scientific insight. As I have mentioned, to understand Archimedes's reasoning in the bathtub or Kekulé's reasoning before his fire, philosophers of science might be helped by cognitive psychology.

The fine granular attention to the primary scientific experimental literature, I believe, sets my work apart from many – but by no means all – studies by New Mechanists. Rather than surveying discovery or justification in whole scientific disciplines or fields, such as biochemistry, cell biology, genetics, neuroscience, or psychology, my case studies focus on the reasoning in the primary scientific literature wherein scientists provide arguments for some specific compositional relations.[37] The discussion of Hodgkin and Huxley's first argument in their 1952 paper regarding the

[36] NB: Saying that the singular compositional abductive account is "more at home" is intentionally vague. A more detailed account will be developed in Chapters 2–4.

[37] My granular approach, thus, differs from that found in Bechtel (2006, 2008) or Darden and Maull (1977). It is closer in approach to Glymour (1980a).

initial inward current is a prime example. It would be impractical for a single human to review any significant portion of an entire scientific discipline or field at the level of detail at which I will be working.[38] Indeed, some scientific topics have so many experimental publications dedicated to them that reviewing even a small fraction of them at the level of detail I adopt would be impractical. Here, I have in mind a fixture of the New Mechanist literature, namely, long-term potentiation (LTP).[39] So, rather than review the evidence for all of neuroscience or all of neuronal function or even all of LTP, I focus on the physiological research on the action potential circa 1950. Rather than review the whole body of justifications for compositional hypotheses concerning vision, I focus on efforts to explain the Hermann grid illusion.

1.4 Explanatory and Abductive Pluralism

Universalism about X in the philosophy of science is the view that there is one universal, static scientific conception of X. Thus, universalism about explanation might hold that there is one universal static conception of explanation. Universalism about abduction might hold that there is one universal static conception of abduction. Pluralism, in contrast, holds that there are many distinct conceptions of X that may vary over time, place, and scientific discipline.

The view of explanation and abduction developed in this book is decidedly pluralist, for which there are precedents. Psillos, Gerhard Schurz, Igor Douven, and Jonah Schupbach have each proposed a disunity to abduction/IBE. Psillos argues that

> IBE should be considered as an inferential genus. The several species of the genus IBE are distinguished, among other things, by plugging assorted conceptions of explanation in the reasoning schema that constitutes the genus. For instance, if the relevant notion of explanation is causal, IBE becomes an inference to the best causal explanation. Or, if the relevant notion of explanation is subsumption under laws, IBE becomes a kind of inference to the best nomological explanation, and so forth. ... Many things can be explanations and the explanatory relation is not fixed and immutable. (Psillos, 2007, p. 444)

[38] See Norton (2021, pp. 256–267) for comments on this topic.

[39] Although it is inessential to my point to assign a precise number of publications, there is this observation: "By 1989, the U.S. National Library of Medicine listed some 312 articles with the term 'long-term potentiation' in the title, and, in the 1990s alone, over 1,000 additional articles have appeared" (Shors & Matzel, 1997, p. 597). Considering the additional articles that have since been published, it would be impractical to try to review all the articles in the whole of this literature.

Schurz proposes that there will not be "*one most general* schema" for abduction, as abductions will depend on details of a specific "abduction scenario" (Schurz, 2008, p. 205). In support of this account, Schurz provides a detailed taxonomy of multiple patterns of abduction. Douven proposes that the phrase "inference to the best explanation" is a mere slogan for similar reasons: "we do best to conceive of IBE as the aforementioned slogan, a slogan that not only permits different fleshings-out, but that may even rationally require different fleshings-out, depending on what our interests are" (Douven, 2017b, p. 23). Schupbach proposes that, "If one is a pluralist about the nature of explanation itself, then varieties of IBE may further multiply, with Inference to the Best Causal-Mechanical Explanation, for example, differing from Inference to the Best Covering Law Explanation, and so on" (Schupbach, 2017, p. 41, fn. 43).

Chapters 2 and 3 will articulate a small fraction of what I take to constitute the explanatory plurality Psillos, et al. refer to. It describes a plurality of singular compositional explanations. Of course, in point of logic, scientific explanatory pluralism does not entail scientific abductive pluralism. In point of logic, there can be scientific explanations that scientists do not use in abductive inferences. What combinations of explanation and abduction scientists use is, as is often said, an empirical matter.

This last point about scientific explanatory pluralism is meant, in the first instance, to be a descriptively accurate account of scientific practice. It does, however, have a secondary purpose. Norton (2021) objects that earlier conceptions of abduction/IBE overlook the diversity of philosophical accounts of explanation.

> Philosophical analysis of explanation has failed even to find a univocal sense of explanation at work in science. Instead, it has found multiple, competing senses of explanation. This multiplicity indicates that the notion is a loose one – an umbrella concept covering several disparate notions. They have no common core. (Norton, 2021, p. 248)

Norton is quite right to point this out, but, as Psillos implies, this is not a conclusion that an abductive approach needs to resist. As I shall argue more concretely in Chapters 2 and 3, there are at least three species of singular compositional explanation. In Chapter 4, and in the case studies, I will argue that there are at least three species of singular compositional abduction. In Chapter 6, I will broach the idea of "singular interlevel abductions," an idea that will be developed in greater detail in Chapter 10.

1.5 The Use of Case Studies

As the use of case studies is such a prominent part of this work, I have to address some of the perennial worries about their use. Joseph Pitt has given the issue a trenchant formulation in what he calls "the dilemma of case studies":

> What do appeals to case studies accomplish? Consider the dilemma: On the one hand, if the case is selected because it exemplifies the philosophical point, then it is not clear that the historical data hasn't been manipulated to fit the point. On the other hand, if one starts with a case study, it is not clear where to go from there – for it is unreasonable to generalize from one case or even two or three. (Pitt, 2001, p. 373)

My response to the dilemma is to tackle it head on. In truth, neither horn of the dilemma is insuperable. Regarding the first horn, it is an *a priori* logical possibility that a case study manipulates the historical data. (The use of "data" in this discussion should not be confused with the use of "data" articulated in Chapters 4–11.) Case studies should not, however, be dismissed simply based on this *a priori* possibility. Clarity regarding possible manipulation of a case must be achieved *a posteriori* by comparing the account with the historical record. Regarding the second horn, it is indeed unclear, *a priori*, how one might generalize from cases. Nevertheless, I provide guidance *a posteriori* on how to generalize. I do this by drawing attention to features of the cases that are likely to be applicable to other cases.

This last point merits elaboration. What ties together the cases of compositional abduction that I examine is the use of intralevel experiments. (I depart from this in Chapters 6 and 10 with the discussion of singular interlevel abduction.) In many cases, scientists manipulate some activity or property instance of an individual and then measure some response of that individual, but do not directly measure or manipulate the activity or property instances thought to underlie the connection between the manipulation and the measurement. Recall the situation with Hodgkin and Huxley's 1952 experiments regarding the sodium hypothesis. Hodgkin and Huxley voltage-clamped axons and measured the consequent total currents. They could perform this intralevel experiment. They did not, however, track the movements of individual ions. Indeed, they could not even track the movements of distinct ion species, such as potassium or sodium. Instead, they could only postulate such ionic movements as they thought would explain the interactions between the voltage clamping and the subsequent currents in different media.

This scientific use of intralevel experiments to inform conclusions about lower level goings on is a guide that suggests that scientists are using compositional abduction. Indeed, at this abstract level, one can see that Tolman had engaged in such reasoning about maze experiments with rats. Tolman, or his colleagues, placed rats in start boxes of mazes and measured the rats' movements through the maze, but he did not measure what was going on in a rat's brain when it was navigating a maze. In interpreting such experiments, Tolman relied on compositional abduction. In the child development literature, the preferential looking paradigm is an intralevel method. It involves having an infant view various sequences of events and measuring their responses, which are then used to make inferences about what is going on in the infant's mind. So, merely starting with a single case study may not, in general, make it clear how to proceed, but one can at least entertain a reasoned hypothesis about how to proceed.

De Regt (2017) takes up Pitt's dilemma, but interprets it differently than I do. After citing the passage above, he writes, "In other words, the philosopher who uses case studies is either guilty of 'cherry-picking' or of overgeneralization" (De Regt, 2017, p. 7). As I read the first horn of Pitt's dilemma, manipulation of data is a matter of interpreting a scientific case in a prejudicial manner. In contrast to data manipulation, I understand "cherry picking" as selecting individual cases, while ignoring similar cases that contradict one's preferred account. One might get the case cited correct, but one ignores relevant near neighbors. Further, as I read the second horn of Pitt's dilemma, it is a matter of not knowing how to generalize, rather than overgeneralizing. Thus, as I read things, there are two dilemmas in play. There is Pitt's dilemma of case studies and de Regt's dilemma of case studies. Both are serious and must be addressed.

As with Pitt's dilemma, it is indeed *a priori* logically possible that the cases are cherry-picked. Whether a case is cherry-picked must be determined *a posteriori* by looking to nearby cases. One precaution I take against cherry picking is to use a case favored by the New Mechanists, the Hodgkin–Huxley theory of the action potential. In this case, there is at least the common ground that scientists are dealing with compositional explanations. That is not all. In Chapter 5, I do not cover just one experiment from Hodgkin and Huxley. I review the interpretation of many experiments in mid-twentieth-century physiology. In Chapter 6, I review several different experiments regarding the biological basis of the Hermann grid illusion. *A priori* one cannot tell if these many examples are cherry-picked. Skeptics will have to review the relevant history for themselves.

As for overgeneralization, my general strategy is to err on the side of caution. My default proposal is to claim merely that there exist cases of a particular description, namely, the ones demonstrated. An existential conclusion is, of course, a weak conclusion, but the existence claims are "beachheads" for further theoretical exploration. Such a cautious default is warranted as many of the proposals I offer barely register on the philosophical landscape. Peirceans rarely, if ever, entertain the possibility of abduction being used for confirmation. Compositional abduction is barely noticed in the IBE literature. New Mechanists have rarely commented on abduction or IBE. Perhaps one can first agree that there are cases as I describe, and then turn to how pervasive they are. That is my strategy.

As an aside, I would add that I accept both Pitt's and de Regt's dilemmas. They are genuine challenges that the historian and philosopher of science must face. De Regt offers a different reply. He proposes to "dissolve" the dilemma on the grounds that it rests "on a misguided view of the relation between philosophy and history of science. It assumes what Jutta Schickore (2011) calls the 'confrontational model of HPS,' on which philosophical theories of science have to be confronted with historical data" (De Regt, 2017, p. 7).[40]

I do not embrace de Regt's dissolution for two reasons. First, for my project, I am very much concerned to draw attention to features of scientific reasoning that have largely been ignored by philosophers of science. One may read this as testing extant philosophical theories against the historical record. This does not commit me to drawing an analogy between this philosophical method and a scientific method of testing theories by experimental data. Nor does this commit me to claiming that confrontation is *the* relation between philosophy of science and the history of science, but it is one important relation. Nor does this commit me to a history and philosophy of science model in which the *discipline* of the history of science provides data for the *discipline* of the philosophy of science. In my experience, historians of science have their own intellectual agendas, which generally do not include being aides to philosophers of science. Historians of science are rarely interested in, say, whether Hodgkin and Huxley used abduction to introduce the sodium hypothesis (a la Peirce), rather than to confirm the sodium hypothesis, as I contend.

Second, de Regt's preferred model of integrated history and philosophy of science does not avoid the dilemma(s). He proposes that

[40] Schickore (2011).

> philosophical analysis and historical research should continuously and mutually interact. . . . The attempt to construct a coherent narrative may require articulation or modification of the employed philosophical concepts, which can subsequently be applied in other historical case studies. (De Regt, 2017, p. 9)

This is a sensible reply. There should indeed be a continual interaction between one's philosophical account and the historical record. Starting with particular historical episodes is reasonable as well. My science-first approach embraces it. I would say that the account of compositional abduction in Chapter 2 constitutes a relatively neat package of ideas, but that close contact with scientific cases precipitated a less organized package in Chapter 3. Indeed, throughout the text I will draw attention to cases where historians and philosophers of science should distinguish singular compositional abductive inferences from other "neighboring" inferences they are often combined with. Such neighboring inferences may be simple enumerative inductive inferences.

Nevertheless, even using an iterative method and starting with a particular historical episode, Pitt's and de Regt's dilemmas remain. One must take care not to distort a case to fit one's preconceptions. One must figure out *a posteriori* how, if at all, one might generalize from a particular historical episode. One must take care not to cherry pick one's examples. One must be careful not to overgeneralize. De Regt does not speak to the matter of how one addresses these concerns with an iterative method, whereas I have at least given it a try.

1.6 Summary

This book offers a theory of how scientists confirm compositional hypotheses. It is driven by a close reading of scientific experimental work. Because it is offering a theory of these practices, it uses philosophical concepts that are often unfamiliar to scientists themselves. To what extent the scientific ontological commitments described here conform to reality is the topic for another project; to what extent the reasoning described here is rational or warranted is also the topic for another project. A psychologically informed account of the "moment of insight" is another important project set aside for future work. While there are no *a priori* guarantees that our case studies are correct and representative of larger patterns in science, this must be assessed largely on an *a posteriori* basis.

CHAPTER 2

Singular Compositional Explanations

If there were such a thing as a complete theory of scientific abduction – a complete theory of all the types of abductive inferences scientists use – it would include a complete theory of all the types of explanations used in scientific abductive inferences. This book develops a fragment of such a hypothetically complete theory. It is a theory of some, but likely not all, scientific abductions. This chapter, therefore, introduces a theory of some, but likely not all, scientific explanations.

There is, of course, a large philosophical literature dedicated to scientific explanation. A theory of scientific compositional explanation alone could constitute a book. However, since this book is on scientific compositional abduction, space must be left for the inference theory. Moreover, space must be left to show that the theory of compositional abduction is, in fact, a descriptively accurate account of at least some actual scientific reasoning. And many philosophers will want to know how the view of compositional abduction developed here relates to a wide range of potentially divergent views. So, there must be attention to that. To accommodate all this, the account of explanation given here will be relatively brief – just two chapters. It will leave aside many qualifications, justifications, and references to the philosophical literature. Because of space limitations, my goal in this chapter is primarily to orient the reader to my view, rather than to persuade the reader of my view. Many of the ideas in this chapter will be relatively familiar – even if objectionable – to those working on the philosophy of scientific explanation. In contrast, the principal ideas in Chapter 3 will be new. They are ideas that are necessary in an account of scientific compositional abduction that is of passably broad scope. They are ideas that arise from the interaction between philosophical theorizing and the examination of historical cases.

The theory developed here begins, in Section 2.1, with a realist conception of explanation, according to which an explanation is a representation of an ontological dependence relation. To flesh out this conception, one needs to specify the entities standing in the ontological dependence relation(s), the ontological dependence relation(s), and the representation of these. Section 2.2 presents the ontological part of this story, whereas Section 2.3 reviews the representational. The entities standing in the ontological dependence relation(s) are spatiotemporal particulars that fall into three ontological categories, namely, individuals, activity instances, and property instances. There are three ontological dependence relations: constitution, implementation, and realization. These entities and their relations need not be represented by arguments, but may rely upon a wide range of representational formats, such as images, sentences, or tables. On this account, there is a genus of explanation, singular compositional explanation, of which there are three species corresponding to the three ontological categories. This constitutes a positive account of a plurality of distinct singular compositional explanations. Section 2.4 argues that singular compositional explanations are members of a genus of explanation.

2.1 A Realist Conception of Explanation

To introduce the realist conception of explanation that I take scientists to embrace, I begin with an empiricist foil, more specifically, an empiricist version of the Deductive Nomological (DN) model of explanation.[1] I add the qualification "an empiricist version" as my expository purposes will be served even if other versions of the DN model do not have the features I note.[2] Further, my expository purposes will be served even if the version does not do justice to the nuances of any particular philosopher's views.[3] To put the matter another way, at this point, I do not intend to do serious history of philosophy. Instead, I am doing theory exposition.

On the empiricist DN model, as I envision it, the explanation of a singular fact or event is a deductive proof of a statement expressing the thing to be explained – an "explanandum statement" – from statements expressing one or more laws and one or more particular facts – the

[1] The phrase "conception of explanation" is borrowed from Salmon (1984b). Bokulich (2018) uses the term in the same way. By contrast, the compositional explanations that I describe in this chapter are what Bokulich would call "accounts of explanation."

[2] Other philosophers have hinted at, or defended, realist versions of the DN-model. For example, Jaegwon Kim comments that "appropriate D-N arguments are particularly perspicuous as a vehicle for conveying causal information" (Kim, 1994, p. 55). Strevens (2008, 2012) offers a theory of explanation in which deductive (and other representative) relations are supposed to mirror causal relations.

[3] Salmon (1984a) sees Hempel as wavering on the "empiricist conception," whereas Kim (1994) sees Hempel as endorsing this version.

"explanans statements." This may be presented in a familiar schema for the model,

$$\frac{\begin{array}{c} C_1, C_2, \ldots, C_m \\ L_1, L_2, \ldots, L_n \end{array}}{E,}$$

where $C_1, C_2, \ldots, C_m$ are the statements of the particular facts, $L_1, L_2, \ldots, L_n$, are law statements, and E is a statement of the fact to be explained. The deductive relation between the premises and the conclusion means that by accepting the truth of the premises and seeing the validity of the deduction, one has reason to accept the truth of the explanandum statement. In other words, the soundness of the argument makes the event to be explained "rationally expectable." One must incorporate further details into the model, but these suffice for present purposes.

This empiricist version of the DN model has two features that are "non-worldly." First, any ontological relation between the entities in the world that are mentioned in the explanans sentences, on the one hand, and in the explanandum sentence, on the other, are set aside in favor of a logical deductive relation between the explanans sentences and the explanandum sentence. In this model, what does the explanatory work, so to speak, is the logical deductive relation. The second "non-worldly" feature of the model is that it makes the event described by the explanandum sentence in some sense rationally expectable. If one accepts the truth of the explanans sentences and recognizes the entailment, then one has some reason to infer the truth of the explanandum sentence. As I understand this model, one of the principal strengths of the idea that explanations are a species of valid deductive arguments is that it enables empiricists to give an account of something that looks plausibly like explanation but avoids metaphysical commitments about ontological dependence relations among things in the world.

"Non-worldly" empiricism slights what I take to be most scientists' interest in what happens in the world. One might have the sentences, "All sodium salts burn orange in a Bunsen burner flame" and "X is a sodium salt" that entail "X burns orange in a Bunsen burner flame," but what scientists seek in at least some explanations is an accounting of how worldly features of X led to its burning orange when placed in the flame of a Bunsen burner. What is it about that sample of sodium salt – that spatiotemporally localizable individual in the world – that made it burn orange? More generally, what in the world "links" the entities mentioned in the explanans sentences to the entities mentioned in the explanandum sentence? Insofar as a version of the DN model denies that something about being a sodium salt makes it burn orange, that version is empiricist. Insofar as a version of the DN model renders an explanandum sentence rationally expectable, rather than specifying what in the world brought about the circumstance to be explained, the version is empiricist.

A realist conception of compositional explanation – in contrast to an empiricist conception – invokes something over and above the logico-linguistic.[4] Explanation is not merely representational, nor is it merely ontological. On a realist conception, a compositional explanation has both representational and ontological components. On a realist conception, a singular compositional explanation is a representation of a real compositional relation between things in the world that are to be explained and things in the world that are to do the explaining.

One of the attractive features of explanatory realism is that it offers a diagnosis of the problems underlying many of the famous counterexamples to the DN model of explanation. As is well-known, one might have a proof of a height-of-a-flagpole sentence that includes a length-of-the-shadow sentence, but this is not an explanation of the height of the flagpole. The realist diagnosis is that the logical relation is not explanatory. It does not do explanatory work. The length-of-the shadow "explanation" of the height of the flagpole does not work, since the length of the shadow does not ontologically determine the height of the flagpole. Further, pressing the point, it is not the logical relation between the height-of-the-flagpole sentence and the length-of-the-shadow sentence that does explanatory work. Instead, the work is done by the ontological dependence of the length of the shadow on the height of the flagpole. Consider another familiar objection to the DN model. Scientists do not explain a man's not becoming pregnant in terms of that man's taking birth control pills. There may be logical relations between sentences, but what is missing is the ontological dependence between taking a man's taking birth control pills and his not becoming pregnant. Another example: the change in the barometer reading does not explain the storm, since the storm does not ontologically depend on the barometer reading.

[4] My terminology borrows from Kim (1994), who developed his view, in part, through an examination of Wesley Salmon's three conceptions of explanation. See, for example, Salmon (1984a, 1984b). Kim's terminology has been adopted by various epistemologists, for example, Taylor (2018), Roski (2021), and McCain (2022), but has been all but ignored by New Mechanists.

I should emphasize here that, consistent with my metatheory described in Chapter 1, my goal is not, strictly speaking, to advance a realist theory of explanation *simplicter*. I do happen to hold this theory, but championing that theory is not my point in this book. Instead, consistent with the metatheory described in Chapter 1, my goal is to describe a realist theory of explanation that I think scientists have sometimes implicitly employed. What I am proposing, then, is that, in the case studies of Chapters 5 and 6, scientists are realists when it comes to explanation. The qualification regarding the case studies is not a mere hedging of bets. It is meant to recognize that some scientists have rejected the idea that science aims to offer explanations. Famously, Karl Pearson claimed that "Nobody believes that science explains anything; we all look upon it as a shorthand description, as an economy of thought" (Pearson, 1911, p. viii).[5] Here, I will take Pearson at his word that he, at least, did not aim to offer scientific explanations. Nevertheless, Hodgkin and Huxley, among others, did accept that scientists offer explanations.

To make this last point in another way, in my case studies, scientific work did not reflect the logical empiricist thinking about explanation that dominated mid-twentieth-century philosophy of science and that has continued to cast a long shadow over the philosophy of science up to the present. Similarly, as I will argue in Chapter 4, scientific work did not reflect logical empiricism's commitment to HD confirmation. Contemporary philosophical theories of how scientists explain and confirm need to move beyond empiricist interpretations of science.

2.2 The Ontology of Compositional Explanations

Once one attributes to a scientist a realist conception of compositional explanation – one that admits roles for both representation and ontology – something must be said about both the representational and ontological. I begin with the ontological in Section 2.2, so that it will be clearer what the representations represent in Section 2.3.

To address the ontological side of things, one may ask a deliberately broad question, "What in the world do scientists think is involved in a compositional explanation?" The answer has two parts. There are (1) the entities in the explanandum and the explanans and (2) the compositional relation or relations between the entities in the explanandum and the explanans. In other words, there are the relata of explanatory relations and

[5] Cited in Salmon (1984a). Also discussed by Oleksowicz (2023).

the compositional relation(s) between them. Section 2.2.1 will describe the proposed relata; Section 2.2.2 will describe the proposed relations.

2.2.1 *The Relata of Compositional Relations*

My theory of compositional explanation focuses on singular compositional explanations. These are compositional explanations scientists sometimes give of spatially localizable, temporally datable particulars. Such explanations include explanandum and explanans entities that fall into at least three distinct ontological categories: individuals, property instances of individuals, and activity instances of individuals. Here, I provide illustrations of each of these ontological categories related to the case study to be presented in Chapter 5.

J. Z. Young is widely credited with the discovery of the squid giant axon, an axon many times larger than other axons known at the time.[6] In short order, this discovery became fundamental to enabling intracellular recordings of axonal activity.[7] Table 1 of Young (1936) reports the number of axons (individuals) counted in nerves (larger individuals) dissected from squid at the Marine Biological Station at Plymouth in the United Kingdom in 1935. Young indicated that various nerves in the dissected squid were made up of these axons. More technically, Young indicated that these axons would provide a singular analytical compositional explanation of what made up the squid nerves. In cases like this, the explanandum and the explanans entities were individuals.

Consider one of Hodgkin and Katz's experiments with intracellular recording. Hodgkin and Katz (1949a) report values for the resting potentials (property instances) of multiple individual squid axons measured at Plymouth, recorded sometime during the summer of 1947. Hodgkin and Katz, in time, concluded that the resting potential of a cell (a property instance of an individual) is explained in terms of the charges on intracellular and extracellular ions (instances of other properties of other individuals).

Hodgkin and Huxley (1952a) illustrate currents in squid axons no. 15 and no. 21 measured at Plymouth during the late summer of 1949. These are multiple activity instances of two individuals. Eventually, Hodgkin and Huxley concluded that these currents are explained by sodium and potassium ions moving across the axonal membrane.

[6] Young (1936). [7] Hodgkin and Huxley (1939), Curtis and Cole (1942).

The preceding examples of individuals, property instances, and activity instances from the work of Young, Hodgkin, Huxley, and Katz are just that, examples. What about some characterization of them? Here, I will offer less than a definition of these items, something less than a set of necessary and sufficient conditions for them. I will, however, draw attention to some of the characteristics of individuals, property instances, and activity instances and indicate some of the ways they are interrelated.

Scientists presuppose that individuals bear property instances. An individual human being may have an instance of normal color vision. An individual molecule of gamma-aminobutyric acid (GABA) will have instances of properties of size, shape, and charge. The property instances of an individual may be manifest or latent relative to some activity. An instance of normal color vision may manifest or reveal itself in human color perception when an individual human takes a color matching test, such as the Farnsworth Munsell 100 hue test or the Ishihara test. By contrast, an instance of normal color vision may be latent relative to seeing colors as when that person is asleep or trying to see in very low light. Instances of the size, shape, and charge of a GABA molecule may manifest themselves in the activity of the molecule's binding to a receptor, but some of the instances may remain latent relative to the activity of binding when the molecule is contained in a synaptic vesicle.

The qualifier "relative to some activity" is meant, in the first place, to capture the fact that property instances can manifest themselves in different ways. So, for example, the dipole moment of a single water molecule – a property instance – might manifest itself in an activity instance of forming a hydrogen bond that raises the boiling point of water. Alternatively, that instance of the dipole moment might manifest itself in an increase in the distance between two molecules of water as they align during the freezing and expanding of water. The qualifier is also meant to avoid a commitment to holding that a property instance is ever absolutely latent, that it does not manifest itself in any activity instance at all. The dipole moment of a water molecule might not manifest itself in an activity instance of freezing or boiling, but there might remain some other activity instance in which it is manifest. Maybe the dipole moment of a molecule always manifests itself in some activity or another.

I have just assumed that scientists presuppose that individuals have property instances and engage in activity instances. Scientists take property instances and activity instances, however, to differ in at least three respects. First, whereas property instances may be latent or manifest, activities are always manifest. Second, activity instances have rates, whereas property instances do not. Third, property instances enable activity instances.

First, whereas property instances may be either manifest or latent, activity instances are always manifest. By "manifest," I do not mean that activity instances are always measured or detected by scientists. Instead, I mean merely that activity instances are active. They are doings. Visually discriminating a numeral in one of the Ishihara plates is an instance of an activity. A capacity for visually discriminating a numeral in one of the Ishihara plates, a property instance, is not the same thing as the activity of discriminating a numeral. Binding to a receptor is an instance of an activity. A capacity for binding is not the same thing as an activity instance of binding.

Property instances differ from activity instances in a second regard. Since an activity instance is a doing, it is something that takes time. Visually discriminating a numeral in an Ishihara plate or binding to a receptor takes time. Activity instances can, therefore, have associated rates. There are rates at which numerals in Ishihara plates can be visually discriminated and rates at which GABA molecules can bind to receptors. Thus, activity instances have rates, whereas property instances do not.

Property instances differ from activity instances in a third respect: property instances enable activity instances.[8] A property instance of S during some temporal interval t_0-t_1 may enable an activity instance of S during t_0-t_1. A human's having an instance of normal color vision enables that individual to discriminate numerals in an Ishihara plate. Instances of the size, shape, and charge of a GABA molecule enable that molecule to bind to receptors.

There are three key features of enabling: (1) it is an intra-individual relation, so it is not a compositional relation, (2) its relata are contemporaneous, so it is not a causal relation and (3) it is an ontological dependence relation. Consider these features in order. If there is a property instance of S, it may enable an activity instance of S. Enabling is an intra-individual relation. Thus, "enabling," as I use it, is not a term for a compositional relation. It is not a relation between an individual, property instance, or activity instance at one level and an individual, property instance, or activity instance at another level. If a property instance enables an activity instance during t_0-t_1, then the property instance and the activity instance occur during the same window of time. The property instance and the activity instance are contemporaneous. This feature indicates that the

[8] See, for example, Craver and Darden (2013, pp. 16f, 78f). Craver and Darden do not provide an account of what they mean by "enabling."

enabling relation is not a causal relation, since causes are taken to temporally precede their effects. Nevertheless, enabling is an ontological dependence relation. Were S not to have property instance F during t_0-t_1, then it would not engage in activity instance Ψ during t_0-t_1. If a water molecule were not polar during t_0-t_1, then it would not engage in forming hydrogen bonds during t_0-t_1. What all this suggests is that there are noncausal, noncompositional "enabling explanations" which are representations of ontological dependence relations. As important and undertheorized as are these explanations, their further investigation goes beyond the focus of this book.

To be clear, as I use the term, "enable" is not supposed to be a term of ordinary language. Instead, it is another technical philosophical term used to describe the relation between some property instances and some activity instances. So, one might say in ordinary English that the extinction of the dinosaurs enabled the rise of mammals, but that is not the sense of "enable" I have in mind. In offering this account of enabling, I do not mean to stipulate how I will use a term. Instead, I am offering a theory of a putative relation between property instances and activity instances. Philosophers of science should feel free to offer reasons to believe that this account is mistaken.

These three features are far short of a complete characterization of how scientists understand the differences between property instances and activity instances. Nevertheless, the features do provide a *prima facie* basis for them. Further, having the distinction between property instances and activity instances provides a tool for a more detailed, accurate description of scientific work.

Contrast this ontological framework of individuals, activity instances, and property instances with the New Mechanist framework. New Mechanists often propose a "dualist" ontology of what they sometimes call "entities" (close to what I mean by "individuals") and "activities."[9] For one thing, my framework is explicitly formulated in terms of spatiotemporal particulars. The choice of "activity instances" and "property instances" over "activities" and "properties" is meant to make this clear. For another, where I recognize three ontological categories, New Mechanists propose two. To preserve their dualism, some New Mechanists may propose to use the term "activity" to cover both activities

[9] Cf., for example, Machamer et al. (2000), Craver (2007), Illari and Williamson (2012), Craver and Darden (2013), Kaiser (2017), and Krickel (2018a). For variants of this dualism, see Bechtel and Abrahamsen (2005) and Bechtel and Richardson (2010).

and properties. This terminological proposal saves the New Mechanist dualism from a simple counterexample. Nevertheless, the goal here is to develop a philosophy of science that provides a descriptively adequate account of scientific compositional abductive reasoning regarding experimental results. To this end, there needs to be some distinction between activities and properties. The New Mechanist who would use "activities" to cover both activities and properties would still need a distinction between two subtypes of "activities." One subtype would correspond to the "activity activities" and the other to the "property activities." The issue for the New Mechanist dualism is not that it falls prey to a counterexample; it is that it needs more theoretical detail. A descriptively adequate philosophy of science needs at least three ontological categories.

Before moving on, let me add a methodological clarification. Sometimes metaphysicians propose "analyses" of these ontological categories. Machamer et al. (2000) mentions philosophical attempts to analyze, define, reduce, or understand the activities of individuals in terms of properties at instants.[10] It also mentions proposals by process metaphysicians to reduce individuals to activities.[11]

Such philosophical analyses are orthogonal to my purposes. Most importantly, as noted in Chapter 1, I am not engaged in metaphysics. I am not developing a theory of the (fundamental) structure of reality. I am developing a theory of what scientists believe exists and how they reason. Further, even within the scope of my efforts to understand scientific compositional abduction, it suffices to say that scientists typically treat these ontological categories as basic. Maybe in some journal article somewhere, some neuroscientist or psychologist offers an analysis, definition, or whatever, of individuals in terms of processes, or whatever. In my case studies, however, physiologists and psychologists treat the ontological categories of individuals, property instances, and activity instances as basic. For my part, I will abide by this.

2.2.2 *Compositional Relations*

Turn now from the relata in scientific compositional explanations to the scientific compositional relations. My argument for thinking that scientists believe in these relations is that this makes sense of the scientific idea that putative lower level entities generate or produce higher level entities. These relations flesh out what realists think is really in the world underlying

[10] Machamer et al. (2000, p. 4). [11] Machamer et al. (2000, p. 5).

Table 2.1 *Scientific compositional relations*

Type of entity	Compositional relation
Individuals	Lower level individuals together *constitute* a higher level individual.
Activity instance	Lower level activity instances together *implement* a higher level activity instance.
Property instance	Lower level property instances together *realize* a higher level property instance.

explanations, but what empiricists deny is really in the world. I will offer only some brief comments indicating some of the features scientists take these relations to have and how these differ from other noncausal relations. Defenders of some of the alternative accounts will find my treatment superficial, but again my goal here is mainly expository rather than polemical.

As broached in the Introduction, I propose that scientists are committed to at least three compositional relations – constitution, implementation, and realization – that obtain between individuals, activity instances, and property instances, respectively. Table 2.1 summarizes these relations. Scientists implicitly assume these relations have seven features: (1) they are many-one relations, (2) they are asymmetric, (3) they are transitive, (4) they are relations of natural dependence, (5) they do not involve the transfer of mass-energy, (6) their relata are contemporaneous, and (7) their relata are not wholly distinct.[12]

As mentioned in Chapter 1, compositional relations are many-one relations. Many sodium ions moving across an axonal membrane implement an axon's inward current. Three instances of the electronegativity of atoms of hydrogen and oxygen realize the polarity of a water molecule. Many photoreceptor cells constitute a retina.

Scientists treat constitution, implementation, and realization as asymmetric relations. This reflects the asymmetry of many scientific compositional explanations. Scientists explain what a molecule is made of by citing its atoms, but not vice versa. Scientists explain the action potential in terms of ion fluxes, but not vice versa. Scientists explain the dipole moment of a

[12] For a longer, slightly different list, see Gillett (2016). Note that, strictly speaking, in the literature, these seven ideas typically take the form of metaphysical claims about composition and causation, rather than, as here, the form of descriptive claims about what scientists take composition and causation to be like.

hydrogen fluoride molecule by appealing to the electronegativity of its constituent atoms, but not vice versa.

Although the asymmetry assumption is widely accepted, it is not universally accepted. Craver has proposed that "Constitutive relevance is symmetrical in a way that etiological (that is, causal) relevance typically is not. . . . all constitutive dependency relationships are bidirectional" (Craver, 2007, p. 153).[13] The symmetry assumption plays an important role in Craver's motivation for his "mutual manipulability" account of the testing for constitutive relevance: "This is the core reason why constitutive relevance should be understood in terms of mutual manipulability rather than in terms of the unidirectional variety" (Craver, 2007, p. 153). Thus, abandoning the symmetry assumption would undercut "the core reason" for *mutual* manipulability. This is a point to which I will return in Chapter 10.

Scientists treat constitution, implementation, and realization as transitive relations. Scientists maintain that protons, electrons, and neutrons constitute atoms and that atoms constitute molecules, so that protons, electrons, and neutrons constitute molecules. Scientists propose that ion fluxes implement an action potential of a cell and that action potentials of cells implement an activity instance of a small neuronal circuit, so that ion fluxes implement an activity instance of a small neuronal circuit. Scientists take the electronegativities of hydrogen and oxygen atoms and the bond angles in an H_2O molecule to realize the polarity of that molecule and they take the polarity of water molecules to realize the buoyancy of an ice cube in water, so that they take the electronegativities of hydrogen and oxygen and the bond angles in an H_2O molecule to realize the buoyancy of an ice cube in water.

A fourth feature of the scientific concepts of constitution, implementation, and realization is that they are natural dependence relations. By this I mean that scientists presuppose that these relations obtain only under certain background conditions, the most familiar of which are temperature and pressure. Squid giant axons only constitute nerves under certain temperatures and pressures. Extreme highs or lows destroy the axons and nerves. As Hodgkin, Huxley, and Katz sometimes noted in passing, resting and action potentials are sensitive to temperature. Again, extreme highs and lows destroy the axon, hence their resting and action potentials. Natural dependence relations, such as constitution, implementation, and realization, differ from "non-natural" dependence relations in this regard.

[13] See Schindler (2013) for a critique of Craver's view. See Krickel (2024, p. 14) for a more recent endorsement of this view.

Logical and semantic entailment, for example, do not hold under background conditions.

A fifth feature of compositional relations, as scientists conceive them, is that they are "mass-energy neutral." Consider again the water molecule. The two hydrogen atoms and the oxygen atom have some mass-energy. When covalently bonded, these atoms form a not wholly distinct entity, a water molecule. This molecule, however, does not have mass-energy over and above the mass-energy of the constituent atoms. (This is not to deny that the bonding of separate atoms releases energy.) Contrast the compositional case with a case of adding an atom. Three hydrogen atoms and an oxygen atom have greater mass-energy than the two hydrogen atoms and the oxygen atom. In this case, the addition of a distinct entity does increase mass-energy.

The compositional relations to which scientists are committed are distinct from the causal relations to which they are committed based on the assumption that causes must precede their effects, whereas composing entities are contemporaneous with the composed entity. This marks a sixth feature of compositional relations. An individual water molecule is contemporaneous with its individual constituent atoms of hydrogen and oxygen being covalently bonded. An individual red cone opsin Red (Ser^{180}) – the human red cone opsin having a serine in position 180 – is contemporaneous with the chain of individual amino acids that constitute it. A property instance of buoyancy in an individual cork is contemporaneous with the property instances of impermeability of the cork's individual constituent cells. A property instance of the dipole moment of an individual water molecule is contemporaneous with the property instances of the electronegativities of the molecule's constituent atoms. An activity instance of an initial inward ionic current of an individual squid giant axon is contemporaneous with the activity instances of individual sodium ions moving into the cell. An activity instance of an individual ion channel opening is contemporaneous with the activity instances of the movement of the channel's individual constituent monomers. Notice that when many instances of x_i ϕ_i-ing implement S Φ-ing from t_0 to t_1, the assumption is not that each of the x_i ϕ_i-ing is engaged in its activity instance throughout t_0-t_1. Instead, it is that from t_0 to t_1 there is some x_i ϕ_i-ing. The influx of sodium ions is a clear illustration. For the duration of the initial inward current from t_0 to t_1, there are sodium ions entering the axon. There is, however, no single sodium ion that spends the entire interval t_0-t_1 crossing the axonal membrane.

To forestall a possible confusion, recall that, back in Section 2.2.1, I proposed that one feature of the enabling relation is that its relata are contemporaneous. During t_0-t_1, a property instance enables some activity instance. In this section, I have proposed that the relata of the compositional relations constitution, implementation, and realization are contemporaneous. This is an empirical hypothesis about how scientists treat these relations. There is nothing mysterious about enabling, constitution, implementation, and realization all involving contemporaneously existing relata. This is no more mysterious than various relations, such as being older than, being heavier than, or being larger than all being transitive relations.

In drawing the preceding distinction between how scientists implicitly understand causation and composition, I assume that I am describing what other philosophers have described in terms of diachronic versus synchronic relations.[14] The idea that causes temporally precede their effects is sometimes described by saying that the causal relation is diachronic. The idea that the composed is contemporaneous with the composing is sometimes described by saying that compositional relations are synchronic.

Some philosophers (critical of the causation/ composition distinction), however, interpret the diachronic/synchronic terminology differently.[15] They assume that the idea is that the diachronicity of causal relations means that this relation exists over time, whereas the synchronicity of compositional relations means that these relations hold only at an instant. They then point out that such a synchronic compositional relation does not correctly describe scientific cases, so that a new conception of compositional relations is needed. While I take such critics of the diachronic/ synchronic distinction to be uncharitable in their reading, I will agree with them that their version of the diachronic/synchronic distinction does not capture the scientific compositional cases. My term "contemporaneous" is meant to capture the fact that compositional relations hold over an interval of time and that the composed and composing stand in this relation.

Consider yet another clarification. Scientists treat many of the relata of compositional relations as dynamic. The relata in a single relation change over time. To take an unproblematic feature, water molecules and proteins oscillate. However, one should not infer, simply on that basis, that scientists think that the compositional relation must also be dynamic. The dynamic character of the *relata* is insufficient basis for thinking that the *relation* is dynamic. This is no more remarkable than the idea that two individuals may remain siblings even as they age.

14 See, for example, Craver and Bechtel (2007), Gillett (2016), and Bennett (2017).

15 For example, Kirchhoff (2013).

There is an important, but underappreciated, implication of the idea that compositional relations are between contemporaneous particulars. In many instances, scientists propose to explain, say, an instance of a rat's behavior by reference to activity instances of parts of the rat's brain, as in the computations in a hippocampal cognitive map. If we take an instance of the rat's behavior to be S Ψ-ing and an instance of the hippocampus's activity as x_i ϕ_i-ing, then strictly speaking the x_i ϕ_i-ing will not be contemporaneous with S Ψ-ing. The behavior instance of S Ψ-ing can begin before the activity instances x_i ϕ_i-ing begin. Maybe the behavior begins with a movement of the rat's head or eyes. Further, the S Ψ-ing can continue after the activity instances x_i ϕ_i-ing cease. Maybe the behavior instance includes contractions in the rat's leg muscles. In notation, for successive times t_0, t_1, t_2, t_3, S Ψ-ing can occur during the interval t_0-t_3, whereas the x_i ϕ_i-ing occurs during the interval t_1-t_2. More succinctly, the explanans x_i ϕ_i-ing will not be contemporaneous with the explanandum S Ψ-ing. *A fortiori* the x_i ϕ_i-ing will not stand in the implementation relation to the S Ψ-ing. How is the scientific practice to be reconciled with the philosophical theory of compositional relations?

Sticking with the case of activity instances, the reconciliation I propose is that scientists accept the x_i ϕ_i-ing as an explanation of the S Ψ-ing if they believe the x_i ϕ_i-ing implement a "contributing" activity instance S Ψ_C-ing for S Ψ-ing. In general, the contributing activity instance S Ψ_C-ing is not named or specified, but it is assumed to be an activity of S and is assumed to be contemporaneous with the x_i ϕ_i-ing. This is a loose description of a contributing activity instance. Probably inadequate as a characterization. Still, the idea is that a hippocampal activity instance explains rat navigation on the assumption that the hippocampal activity compositionally explains a contributing activity instance of the navigation. If one were to add some terminology, one might say that scientists accept a "derivative compositional explanation" of S Ψ-ing in terms of x_i ϕ_i-ing, if there is a compositional explanation of S Ψ_C-ing in terms of x_i ϕ_i-ing.

Consider another example of the importance of contributing activity instances. In Chapter 6 we shall see that scientists often look to activity instances of retinal ganglion cells (RGCs) as offering an explanation of instances of the Hermann grid illusion. RGCs do not compositionally explain the *experience* of the illusion. Assuming that the experience is implemented in the cortex, RGCs do not explain the experience, since they are found in the retina. Instead, what scientists have in mind is that RGCs offer a derivative explanation of the *perception* of the illusion, where an activity instance of perception is implemented by activity instances in the retina, the optic tract, and various regions of the cortex. The RGC activities offer a derivative compositional explanation of an instance of the perception of the Hermann grid illusion because they provide a compositional explanation of an activity instance S Ψ_C-ing that contributes to the perception of the illusion. In this regard, the RGC example differs from another example to be discussed in Chapter 10. In Chapter 10, I will review an example wherein certain cortical cells are thought to implement an experience of a sensation of flutter. Many examples of compositional explanation in cognitive science are plausibly interpreted as "derivative compositional explanations."

The Hodgkin–Huxley theory of the currents associated with the action potential provides a valuable contrast with the foregoing cases. One of Hodgkin and Huxley's many achievements was to recognize the total ionic current as the sum of a capacity current, a sodium current, a potassium current, and a leak current. Hodgkin and Huxley developed a compositional explanation of the capacity current in terms of the movements of charge up to the axonal membrane. Further, they developed compositional explanations of each of the ionic currents in terms of movements of ions across the axonal membrane. Each current, however, is a contributing activity instance of an instance of a total current. In other words, part of what the equation $I_{tot} = I_C + I_{Na} + I_K + I_{leak}$ tells us is that an instance of a capacity current, I_C, an instance of a sodium current, I_{Na}, an instance of a potassium current, I_K, and an instance of a leak current, I_{leak}, are contributors to a total current instance, I_{tot}. So, one might say, informally, that sodium influx explains part of the action potential. More formally, one might say that sodium influx compositionally explains the sodium current, the potassium efflux compositionally explains the potassium current, . . ., which explains the action potential.

Finally, the seventh feature of the scientific conception of compositional relations is that their relata are not wholly distinct. In a causal relation, the individual associated with the cause is wholly distinct from the individuals associated with the effect.[16] In the New Mechanist literature, Craver and Bechtel defer to David Lewis:

> C and E must be distinct events – and distinct not only in the sense of nonidentity but also in the sense of nonoverlap and nonimplication. It won't do to say that my speaking this sentence causes my speaking this sentence or that my speaking the whole of it causes my speaking the first

[16] Cf., for example, Lewis (2004) and Craver and Bechtel (2007).

half of it; or that my speaking causes my speaking it loudly, or vice versa. (Lewis, 2004, p. 78; cited in Craver & Bechtel, 2007, p. 552)

Compositional relations allow the violation of the nonoverlap condition. An individual water molecule "overlaps" with its hydrogen and water atoms. An individual cork having a property instance of being buoyant overlaps with its individual constituent cells having property instances of impermeability to water. An individual ion channel engaged in an activity instance of opening overlaps with its individual constituent protein monomers engaged in their activity instances of moving.

Before moving on I should note one feature that does not make the list: spatial containment.[17] It is sometimes said that, in a compositional explanation of an activity instance of an individual, the explanans individuals are spatially contained in the explanandum individual. Accordingly, the compositional relation should have the individuals cited in the relata of the explanans be spatially contained in the individual cited in the relatum of the explanandum. Although the spatial containment condition is often satisfied, it is clearly violated in the case of the axonal currents during the voltage clamp. The initial inward sodium current is ontologically determined by the influx of sodium ions into the axon and is obviously a case of individuals not spatially contained in the axon entering the axon. The late outward current is ontologically determined by the efflux of potassium ions and is obviously a case of individuals spatially contained in the axon leaving the axon. The same story might be told in terms of the action potential, rather than axonal currents. Moreover, there are large classes of cases where the spatial containment relation is not respected. When an individual melts, the components of that individual cease to be spatially contained in the individual. When a liquid freezes to form an individual crystal, individuals that are not spatially contained in the crystal come to be spatially contained in the crystal. There is a parallel story to be told in terms of sublimation and desublimation. It is because of violations of the spatial containment condition that I sometimes refer to "associated" parts, where I use "associated" as a black box term that is meant to encompass the cases of nonspatially contained parts.

One might, of course, try to save the spatial containment requirement. One might postulate that, in cases of the action potential, melting, freezing, sublimating, desublimating, and so forth, the explanandum individual is a system and the explanans individuals are spatially contained

[17] See, for example, Krickel (2020, 2024).

parts of this system. While that would technically work, it would be *ad hoc*. What is the evidence that, say, Hodgkin and Huxley drew a distinction between the axon and a larger "axonal system"? Are there instances where scientists postulate ice cubes along with "ice cube systems?"

At various points in my discussion I have referred, and will refer, to a notion of "levels." The topic of levels is the subject of its own vigorous philosophical debate, most of which I will avoid as tangential to my concerns.[18] Pre-theoretically, I assume that an organism's brain is at a lower level than the organism, that an organism's cells are at a lower level than the organism, and that ions are at a lower level than a cell. These assumptions will cover what is needed for the cases studies in Chapters 5 and 6. Further, these are assessments that, *prima facie*, a theory of levels should capture. Although I will not defend a theory of these levels here, if pressed, I would turn to a version of Gillett's "Integrative Account of Levels."[19]

To this point in the chapter, I have been developing a realist conception of explanation that I take to be implicit in scientific work. According to this conception, an explanation is a representation of a compositional ontological dependence relation between entities in at least three distinct ontological categories. Something like this conception can be found in diverse subdisciplines of philosophy, albeit with distinct commitments regarding the ontological dependence relation. By and large, my commitments regarding the ontological dependence relations stems from my focus on the scientific use of singular compositional explanations and the distinctive features of those explanations. If historians and philosophers of science are to have a descriptively adequate account of how scientists compositionally abductively reason, they will need a descriptively adequate account of the compositional relations that scientists presuppose.

My relatively narrow interest in scientific compositional explanations differs from the relatively broad interest of epistemologists who have adopted explanatory realism. John Greco, for example, believes that to understand that p is to have an explanation of p, but to have an explanation of p is to locate p in a system of appropriate dependence relations.[20] Greco's list of dependence relations includes "logical relations, mathematical relations, and various kinds of supervenience relation" (Greco, 2014, p. 286). Maybe one can read Greco's "various kinds of supervenience relation" as covering the compositional relations of constitution, implementation, and realization, but I suppose that logical and mathematical relations are not the dependence relations scientists invoke in singular compositional

[18] See, for example, Brooks et al. (2021). [19] See Gillett (2021). [20] Greco (2014).

explanations.[21] They are not what I described above as natural dependence relations, that is, relations that hold subject to background conditions. My point is that Greco's interest in explanation is broader than my own.

My relatively narrow interest in scientific compositional relations also differs from the broad interest of analytic metaphysicians thinking about ontological dependence and Ground. The introduction to Thompson (2022) provides a helpful orientation:

> The notion that ontological dependence is intended to capture is that of a distinctively metaphysical dependence relation. Common examples of ontological dependence proposed in the literature include the following: a nonempty set ontologically depends on its members; the holes in a piece of cheese ontologically depend on the cheese itself; normative facts ontologically depend on nonnormative facts; and mental phenomena ontologically depend on physical phenomena. Ontological dependence is usually thought to help provide for a distinctive kind of metaphysical explanation. (Thompson, 2022, p. 1)

Clearly, the analytic metaphysicians' project casts a broad net. The case of mental phenomena ontologically depending on physical phenomena probably covers the singular compositional explanation of mental phenomena in terms of physical phenomena, although possibly a lot more. Maybe it would cover the explanation of putative psychological laws in terms of physical or neurophysiological laws. My account of singular compositional explanation is not meant to address relations between laws. Thompson's other examples, however, are completely outside the scope of what I take to be singular compositional explanations.

Thompson's outline captures the spirit of earlier discussions of Ground. Fabrice Correia and Benjamin Schnieder provide a longer list of examples of putative instances of Grounding:

1. Mental facts obtain because of neurophysiological facts.
2. Legal facts are grounded in non-legal, e.g. social, facts.
3. Normative facts are based on natural facts.
4. Meaning is due to non-semantic facts.
5. Dispositional properties are possessed in virtue of categorical properties.
6. What accounts for the existence of a whole is the existence and arrangement of its parts.

[21] See also Grimm (2014), who traces the idea to Kim (1994).

7. A set of things is less fundamental than its members.
8. What makes something beautiful are certain facts about the reception of its beholders.
9. A substance is prior to its tropes or modes.
10. That snow is white is true because snow is white. (Correia & Schnieder, 2012, p. 1)

As with the Thompson list, Correia and Schnieder's item 1 probably includes more than singular compositional explanations. Item 6 is very close to the idea of providing what I call an "analytic compositional explanation" of an individual. Nevertheless, many, if not all, of the other cases plausibly fall outside the scope of a theory of singular compositional explanation.

One more illustration. Michael Clark and David Liggins provide their own list of the putative instances of Grounding:

(A) The brittleness of the cup results from the way its constituent atoms are arranged.
(B) The truth-value of a proposition is determined by how the world is.
(C) Actions have their moral properties in virtue of their non-moral properties.
(D) Non-empty sets depend for their existence on their members.
(E) A mental state is grounded in the brain state which realizes it. (Clark & Liggins, 2012, p. 812)

In this list, it is only (A) and (E) that plausibly include instances of a singular compositional relation to be invoked in a singular compositional explanation. The analytic metaphysicians, however, are implicitly interested in cases such as the single set {Socrates} ontologically depending on the individual human Socrates. See Thompson's first example, Correia and Schnieder's 7, and Clark and Liggins' (D). My account of compositional relations does not aspire to characterize this case.

The examples of work in analytic metaphysics serve three important functions. First, it provides an opportunity for me to reiterate the point from Chapter 1 that this book, unlike the work of the analytic metaphysicians, is not a work of metaphysics. This book is not concerned with the fundamental structure of the world. It is not even concerned with the structure of the world. Instead, it focuses on how scientists *think* the world is structured. It is, for the space of this book, an open question whether the world is, in fact, structured the way scientists think it is structured.

Second, the scope of the aspirations of the analytic metaphysicians is greater than the scope of my aspirations.[22] As I repeatedly noted above, I am focused on singular compositional explanations. I now want to add a bit to that to reiterate another metatheoretical point from Chapter 1, namely, that I do not take singular compositional explanations to be universal. Some of the features of these explanations have emerged in scientific research only in the course of the twentieth century and may well not reflect scientific thinking in earlier eras.[23] So, for one thing, I have proposed that compositional relations reflect mass-energy neutrality. Such a scientific presupposition, however, could only come after a scientific embrace of mass-energy. Such a presupposition could only implausibly be attributed to scientists such as Galileo, Descartes, or Newton. For another thing, I have proposed that singular compositional explanations are used in scientific compositional abductive reasoning, which I take to be used in the scientific experimental literature. However, the nature and organization of the scientific experimental literature is the product of multiple social forces coming to the fore in the twentieth century. It is an understatement to say that this scientific literature is, in certain respects, different than the scientific literature produced by Galileo, Descartes, Newton, and Darwin.

My third point draws together some consequences of the last two. One can now see how it is possible to abandon (would-be) empiricist conceptions of scientific compositional explanation without adopting what many philosophers of science would dismissively describe as analytic metaphysics. I am not doing (analytic) metaphysics since I am not describing the (fundamental) structure of the world. Nor am I engaged in the (analytic) metaphysician's ambitious project of a sweeping account of ontological dependence. My project has a much finer, granular focus on singular compositional explanations that scientists have given at specific times and places. Developing a descriptively adequate account of singular compositional explanations in the recent scientific literature involves granular attention to the arguments scientists give in that literature. This project is, thus, what many would take to be mainstream history and philosophy of science. It may be a project of no interest to epistemologists or analytic metaphysicians.

22 In a well-known critique of Grounding, Wilson (2014) makes the case that there is no work to be done by a general "big G" notion of Grounding. Instead, the work of Grounding might be done in different contexts by multiple "small g" conceptions of grounding. In this context, one might view the compositional relations described in this book as "small g" grounding relations, although not necessarily small g grounding relations that Wilson would embrace.

23 Recall, again, the proposal in Pearson (1911) that science is not engaged in explanation.

2.3 Representation and Compositional Relations

The empiricist version of the DN model described in Section 2.1 proposed that explanations are merely representations. More specifically, they are representations of a particular form, namely, formally valid deductive arguments. Although this version of the DN model has been widely rejected by philosophers over the last forty years or so, the idea that causal explanations and causal relations are aptly represented by DN arguments has been more enduring. For example, Jaegwon Kim commented that "appropriate D-N arguments are particularly perspicuous as a vehicle for conveying causal information" (Kim, 1994, p. 55). Kim does not explain what he means by saying that DN arguments are a "particularly perspicuous" way to convey causal information. Nor does he give any reason to believe that they are. Nevertheless, it is safe to say that scientists rarely use instances of the DN schema to convey causal information. Moreover, it is undoubtedly an understatement to say that in recent years both philosophers and scientists have more commonly used directed acyclic graphs rather than instances of the DN schema to convey causal information.[24] As I look at actual scientific compositional explanations in the special sciences, there is no assumption that these explanations are represented as arguments even superficially looking like the DN schema.

Indeed, once one sets aside empiricist preconceptions stemming from the DN model, one finds that scientists use a wide range of representational devices to communicate both the explanantia and explananda of compositional explanations. These representational devices probably also play a role in the scientific development of compositional explanations, but I set that aside as orthogonal to their role in arguments that are published in the scientific literature. One could probably develop a complicated, multidimensional matrix in which to place different examples. One dimension might be along the develop/communicate distinction, another might be along the explanans/explanandum distinction, and yet another might be along the individual/property instance/activity instance distinction. I will assume that such a complicated system is unnecessary. I will instead review some examples.

In truth, I foreshadowed some of this representational diversity in the examples in Section 2.1. I noted that Young (1936) used a table of cell counts to articulate the explanans of an analytical compositional explanation of what certain squid nerves are made of. I also noted that Hodgkin

[24] See, for example, Spirtes et al. (2000) and Pearl (2009).

and Katz (1949a) used a table to represent values for resting potentials. This table communicated results to be explained. Finally, I noted that Hodgkin and Huxley (1952a) used images of total currents to represent explananda.

It should be unsurprising that different phenomena are represented differently. Scientists use different words, tables, and images to represent different things. Nor should it be surprising to find multiple representations of the same explanandum. Different sentences in the same or different natural languages might refer to the same ontological configuration in an explanation. Here is a more "scientific" gloss of the same point. Conventional differences among representations make for the simplest cases. Physiologists and neuroscientists have used the Nernst equation as a representation of part of an explanans for the resting potential of the axon. Hodgkin et al. (1952, p. 132) and Hodgkin and Huxley (1952a, p. 454) used a version of the Nernst equation expressed as a ratio of an internal ion concentration over the external ion concentration. By contrast, contemporary conventions, as illustrated in Shepherd (1988, p. 91), Hodgkin (1994, p. 273), and Hammond (2014, p. 34), use a ratio of external ion concentration over the internal ion concentration. These mere conventional differences between representations constitute clear (but, in some sense, trivial?) differences among representations of the same ontological thing. Perhaps surprising to philosophers, in recent years neuroscientists have been concerned with helping students and their fellow scientists appreciate the work of Hodgkin, Huxley, and Katz by drawing attention to the (trivial?) differences between mid-twentieth-century representational conventions and twenty-first-century representational conventions.[25]

2.4 Singular Compositional Explanations as a Genus

Singular compositional explanations are a genus of explanation that represents compositional relations among entities. Each species of this genus corresponds to a distinct compositional relation among the entities in a distinct ontological category. I have now added further details to this sketch. I have added some discussion of the representations, the compositional relations, and the entities and their ontological categories. While this is undoubtedly not a complete account, it is at least a preliminary exposition of the principal features of the account.

[25] For further discussion of this and other examples, see Brown (2019, 2020), Hopper et al. (2022), and Raman and Ferster (2022).

This account complements the work of Aizawa and Gillett (2019). That paper proposed that scientists offer three species of compositional explanations: analytic compositional explanations, dynamic compositional explanations, and standing compositional explanations (see Table 0.1). Further, I proposed that each of these species of explanations is backed by a distinct compositional relation, namely, constitution, implementation, and realization, respectively. The present account now articulates what it is for those *prima facie* scientific explanations to be scientific explanations, not in the sense of being "real" scientific explanations but in the sense of being what scientists, in the relevant cases, understand by being an explanation. I have an account of what scientists think they are doing in giving singular compositional explanations.

For those familiar with New Mechanism, what I am calling a singular dynamic compositional explanation is at least similar to what some of the New Mechanists have called "mechanistic constitutive explanation." It may be as similar to existing New Mechanist accounts as those accounts are to each other.[26] However, for those who have not read (Aizawa & Gillett, 2019), the idea of singular standing compositional explanation and singular analytic compositional explanation is likely unfamiliar. So, here I want to articulate, for a New Mechanist audience, some reasons to accept these unfamiliar compositional explanations.[27]

One can make the case for singular standing compositional explanations and singular analytic compositional explanations by providing examples. This is done in Aizawa and Gillett (2019). Take analytic compositional explanations. Aizawa and Gillett (2019) proposed that, during the Scientific Revolution, Robert Hooke explained what corks were made of, namely, cells. Moreover, he explained the composition of the drone fly eye: it is a compound eye made of smaller units that have come to be called "ommatidia." In the nineteenth century, Michael Faraday used distillation to isolate a liquid, benzene, he proposed was made up of molecules of C_6H_6. In the twentieth century, organic chemists learned to use mass spectrometry to identify what makes up various compounds, such as

[26] For some of the variants on the New Mechanist accounts, see Bechtel and Abrahamsen (2005) and Craver (2007).

[27] As far as I can tell, there are few discussions of compositional explanation in the abduction/IBE literature. Certainly, the idea does not appear in some of the more prominent recent works in the area. See, for example, Magnani (2009, 2017), McCain and Poston (2017a), Niiniluoto (2018), and Douven (2022). Lipton (2003, pp. 16, 66–68) mentions "vertical" inferences in passing. "Micro-part abduction" in Schurz (2008) is unlike compositional abduction as understood here, since "micro-part" abduction postulates the same thing, only smaller, whereas in compositional abduction, as understood here, the parts are qualitatively distinct from the wholes.

pheromones. There are also explanations of the composition of anatomical structures, such as the composition of a eukaryotic cell.

Consider some of the standing compositional explanations from Aizawa and Gillett (2019). Hooke explained the buoyancy of a cork in terms of the impermeability of its cells and a drone fly eye's being circumspect in terms of the positions and orientations of the ommatidia. Contemporary biochemists explain the properties of individual proteins in terms of their constituent amino acids, their property instances, and their bonding relation instances. Biologists explain the shape of individual micelles in terms of the property instances of their constituent phospholipid molecules.

In addition to the argument by example from Aizawa and Gillett (2019), I now have a theory that articulates what unifies compositional explanations as compositional explanations. Each of these explanations is a representation of an ontological dependence relation among entities, where each of the ontological dependence relations – constitution, implementation, and realization – shares a family of features. As described in Section 2.2.2, these relations have seven features: (1) they are many-one, (2) they are asymmetric, (3) they are transitive, (4) they are relations of natural dependence, (5) they do not involve the transfer of mass-energy, (6) the relata are contemporaneous, and (7) the relata are not wholly distinct. This is an argument from theoretical unity.

Consider, now, a third somewhat different argument for the same conclusion. To begin, I suppose that scientists in the special sciences postulate individuals, activity instances, and property instances. In the special sciences, scientists postulate individuals, such as atoms, molecules, cells, tissues, and organs. In the special sciences, scientists also postulate instances of activities of these individuals, such as bonding, attracting, adhering, synthesizing, metabolizing, and secreting. Most New Mechanists accept these claims, which one might independently document by a review of the scientific literature. Finally, scientists also postulate property instances, such as charge, polarity, shape, and spectral absorption.

If the scientific literature shows that scientists postulate individuals, property instances, and activity instances in the special sciences, then where do scientists think these entities come from? In principle, scientists might think that they are basic features of the world that have their own independent existence. That, however, is not the New Mechanist take on the science when it comes to activity instances. The New Mechanist's answer for activity instances is that activity instances of parts explain activity instances of wholes. But what of individuals and their property instances? The answer offered here is that there are parallel accounts. Just as there are compositional explanations of activity instances of wholes, there are compositional explanations of individuals and their property instances.

So, there are three reasons to think that there is a genus of compositional explanation containing three distinct species. But what reason is there to resist this conclusion? How might one the resist compositional explanation of individuals? How might one resist the compositional explanation of property instances? Consider two answers.

One path of resistance to compositional explanations of whole individuals is to argue that the sample narratives regarding cells, etc., are not really explanations. Here, the most likely source of resistance is to claim that cells are related to organelles by definition and that organs are related to tissues by definition. One might take as a model the idea that water is defined as H_2O. One might think that it is a conceptual impossibility that water is not H_2O.

To challenge this view, one might first begin with the observation that, even if some philosophers define water as H_2O or find it to be conceptually impossible for water to be other than H_2O, that does not show that scientists have that definition or concept of H_2O.[28] Next, one might look to the philosophy of chemistry literature, where there is some debate about what water is.[29] Part of the concern is that even a sample of "pure" water is not merely a collection of H_2O molecules. H_2O molecules in relatively close proximity to each other will dynamically interact. For one thing, two nearby molecules of H_2O may dissociate to form a hydronium ion (H_3O^+) and a hydroxide ion (OH^-). For another, given the polarity of H_2O molecules, they are prone to forming short chains of differing lengths with the positive side of one molecule hydrogen-bonded to the negative side of the next molecule. Water molecules may not polymerize, but they do oligomerize. The account advanced here has it that scientists think that congeries of individual H_2O molecules, hydronium ions, hydroxide ions, and various H_2O oligomers constitute samples of water. This is why aggregates of H_2O molecules, hydronium ions, hydroxide ions, and various H_2O oligomers explain what water is made of. Similarly, it is why one can explain that organs are made of tissues, that tissues are made of cells, and that cells are made of organelles.

How might one resist the compositional explanation of property instances? One might claim that scientists are not committed to property

[28] See Hendry (2006, 2023) for a similar view of the chemical elements. See Needham (2008) for a defense of the definitional view.

[29] Cf., for example, Needham (2000) and Hendry (2006).

instances. But why think scientists are not so committed? It would be beside the point to argue for the metaphysical conclusion that there are no property instances. Scientists might be committed to property instances, even if metaphysicians were to prove that property instances do not exist. That would be a case where metaphysicians prove the scientists are mistaken about ontology. It would also be beside the point to note that philosophers have not been able to develop any widely accepted theory of what properties are, whether universals, tropes, powers, or what have you. Surely, there are any number of concepts in scientific use that have yet to receive widely accepted characterizations by philosophers. Simplicity, natural kinds, causation, natural law, and explanation are obvious examples.

One might dismiss properties and property instances on the grounds that they are covered by the term "activities." Properties are just a type of activity. I mentioned this idea above in Section 2.1. Unfortunately, this is a mere terminological proposal that papers over the ontological distinction scientists draw between property instances and activity instances. One could call both property instances and activity instances "activities" or "activity instances," but there remain at least three noteworthy differences between the way scientists treat properties and the way they treat activities. First, scientists allow that properties are sometimes latent relative to a given activity. Second, scientists allow that properties may manifest themselves in different activities at different times. Third, scientists propose that properties do not have rates, whereas activities do.

On balance, there are reasons to think that scientists recognize individuals, property instances, and activity instances. Moreover, there are reasons to think that scientists give compositional explanations of some of them. Simply look at scientific practice. Perhaps there are reasons to resist this picture, but as this picture has not been examined in the philosophical literature it is unclear what this resistance might be. The reasons to resist a pluralism of compositional explanations scouted above are not compelling.

2.5 Summary

In this chapter, I have developed an account of singular compositional explanation as a genus of explanation with three species: singular analytic compositional explanation, singular dynamic compositional explanation, and singular standing compositional explanation. Such explanations are representations of a compositional ontological dependence relation among entities in three distinct ontological categories. The representations in these explanations may be quite diverse, relying upon images, tables, and texts, hence displaying a broad range of representation resources. Beneath the representational diversity, however, there is a relatively simple unity to be seen in the common features of the ontological relations. The relatively simple theoretical unity of these examples will not carry over to other scientific compositional explanations to be discussed in Chapter 3.

CHAPTER 3

Singular Compositional Explanations
Special Cases

The core methodological proposal from Chapter 1 is that, if historians and philosophers of science are to understand how scientists try to confirm compositional hypotheses, there is value in looking closely at the reasons they present in the recent experimental scientific literature. This chapter begins to show some of the consequences of this methodological proposal. More specifically, it shows that scientific explanatory practice is more complicated than is captured by the relatively simple theory of singular compositional explanation sketched in Chapter 2. A descriptively accurate account of scientific explanatory practices has to include more details.

Here, I do not aspire to complete a list of scientific compositional explanations. Instead, I provide the bare minimum for addressing the range of compositional abductive inferences I think are to be found in the case studies in Chapters 5 and 6. Here, I envision three additions. Section 3.1 discusses the explanation of rates of activities of wholes. Section 3.2 considers the explanation of experimental results. The point of this section is to connect the theory of singular compositional explanation introduced in Chapter 2 to the context of experimental results. Consequently, one should see why the theory of singular compositional explanation is the way it is: it is to articulate what is sometimes involved in the explanation of results. Section 3.3 will introduce the idea of explaining the results of controlled experiments. Section 3.4 will draw together the elements from this chapter in an illustration of the explanation of controlled experiments involving rates of activity instances.

3.1 Explaining Rates

In Chapter 2, I proposed that one difference between activity instances and property instances is that activity instances take place at rates, whereas property instances do not. Based on this feature, a neuron's resting potential counts as a property instance since it has no rate. In contrast, the action potential at a time has a specific value, say, -60 mV, but it also has a rate of change. Based on this feature, the action potential counts as an activity instance. Of course, at maxima and minima, the action potential has a rate of change of 0 mV/msec, but that is still a rate. Further, to be clear, the resting potential can change, and the change (either at an instant or over some interval) will have a rate, say, +1 mV/msec. However, a change in a resting potential is an activity instance, so it should have a rate. So, there must be a distinction between a resting potential and a change in a resting potential.

Given that activity instances have rates, one might expect that scientists sometimes take an interest in explaining those rates. They want to explain why things happen at the rate they do. Consider, for example, exponential growth and decay. Why did this population of bacteria increase in number exponentially? Because the individual bacteria reproduced in proportion to their numbers. Why did the radioactivity of this sample of U^{238} decrease exponentially? Because the individual atoms decayed in proportion to their numbers. Linear growth and decay occur when the same number of occurrences takes place per unit time. Why did the number of rings in this tree increase linearly? The tree added one ring per year. Why did the soil along this river erode linearly? Each year the same amount of soil was removed.

Of course, not all explanations of rates are grouped by their rates. Consider a more specific and concrete example. In a discussion of an experimental result in 1949, Hodgkin and Katz implicitly asked themselves, "Why did the action potential of this axon rise at the rate it did?" They answered that, "Our hypothesis therefore suggests that the rate of rise of the action potential should be determined by the rate of entry of sodium" (Hodgkin & Katz, 1949a, p. 55).[1] They believed they could explain the rate of rise of the action potential in terms of the number of sodium ions entering the axon. I will examine this example in greater detail in Section 3.4. Chapter 5 will contextualize this example with further details of Hodgkin and Katz's paper that reveal that they were very concerned with the rates at which the action potential rises and falls.

Consider further examples of the explanation of rates far removed from neuroscience. There is an enormous literature on multiple aspects of human aging that focuses on rates of decline. For example, exercise scientists are very interested in changes in performance and rates of change

[1] See Brown (2020, p. 28f) for additional exposition.

in performance with aging.[2] Regarding track and field performance Baker et al. (2003) report that

> There were significant differences in the rates of percentage decline in running events over various distances for both males and females ... There were no differences in rates of declining function for any of the walking events ... The strength-dependent throwing events and pole vault showed the greatest rates of decline with age. (Baker et al., 2003, p. 47)

Given the many examples of scientific efforts to explain the rates of activities, it is surprising that there has been so little New Mechanist attention to the topic.

There is much to be said about the explanation of rates. For present purposes, however, I will only introduce one set of explanations of the rates of activity instances. I believe there are others. The proposal here is that scientists sometimes explain the rate of some activity instance of an individual – the rate of an S Ψ-ing – by appealing to the number of activity instances of its parts – the number of x_i's φ_i-ing. Thus, Hodgkin and Katz explained the initial rate of rise of the action potential in terms of the numbers of sodium ions entering the axon. In another example to be discussed in Chapter 5, Richard Keynes explained the exponential decay in the radioactivity of axons in terms of the number of radioactively labeled potassium ions diffusing out of the axon.

Note that this account of the explanation of rates meshes well with the proposal that scientists sometimes take an interest in spatiotemporal particulars. If Hodgkin and Huxley explained the rate of rise of the action potential in terms of the number of sodium ions entering the axon, then they must have been interested in the individual sodium ions entering the axon. This interest does not, of course, take the form of their being concerned with the individual identities of the individual sodium ions. Instead, the idea is that they had to have some interest in the number of individual sodium ions moving into the axon.

3.2 Interpreting Experimental Results

Sometimes scientists write about the interpretation of experimental results. Sometimes they write about what experimental results mean or what is going on in an experiment. Here, I wish to provide an account of these

[2] See, for example, Moore (1975), Schulz and Curnow (1988), Hollings et al. (2014), and Haugen et al. (2018).

informal locutions. I characterize what scientists are sometimes doing when they interpret experimental results: they are giving singular compositional explanations.

To begin, I distinguish between data and results. Borrowing from James Bogen and James Woodward, I propose that data are what register on measuring devices, such as oscilloscopes.[3] By contrast, results are things in the world scientists are trying to measure, such as action potentials or axonal currents. The term "data targets" might provide a more helpful description of what I have in mind than does "results," but at least some scientists sometimes use "results" in the way I propose. Results differ from what Bogen and Woodward mean by "phenomena" insofar as the former, but not the latter, are spatiotemporal particulars.

To flesh out this distinction, let me begin with a more detailed exposition of what Bogen and Woodward mean by "data." They write,

> As a rough approximation, data are what registers [sic] on a measurement or recording device in a form which is accessible to the human perceptual system, and to public inspection. (Woodward, 1989, p. 394)

> Examples of data include bubble chamber photographs, patterns of discharge in electronic particle detectors and records of reaction times and error rates in various psychological experiments. (Bogen & Woodward, 1988, p. 306, cf., pp. 311, 316–317)[4]

Many mid-century oscilloscope recordings of axonal activity instances fit nicely within the characterization Bogen and Woodward offer. Moreover, the example of oscilloscope traces is on par with the example of bubble chamber photographs. Both of these present measurements or recordings that are accessible to public inspection and to the human perceptual system.

Bogen and Woodward also propose that "Data are . . . idiosyncratic to particular experimental contexts, and typically cannot occur outside of those contexts" (Bogen & Woodward, 1988, p. 317). Further, "many different sorts of causal factors play a role in the production of any given bit of data, and the characteristics of such items are heavily dependent on the peculiarities of the particular experimental design, detection device, or data-gathering procedures an investigator employs. Data are, as we shall say, idiosyncratic to particular experimental contexts, and typically cannot occur outside of those contexts" (Bogen & Woodward, 1988, p. 317).

[3] Bogen and Woodward (1988). [4] See also Woodward (1989, p. 394).

Research on the action potential nicely illustrates the idiosyncrasy of data. Hodgkin and Huxley (1939) reported on experiments performed at the Laboratory of the Marine Biological Association in Plymouth, UK. Their experiments generated data on the action potentials in giant axons isolated from *Loligo forbesi* recorded using a microelectrode made from a glass tube, filled with seawater, to which was attached a silver wire coated with silver chloride at the tip. Kenneth Cole and Howard Curtis performed somewhat different experiments at the Marine Biological Laboratory at Woods Hole, MA.[5] Their experiments generated data on the action potential in giant axons isolated from *Loligo pealii* recorded using pulled glass microelectrodes filled with KCl isosmotic with seawater. Hodgkin and Huxley used one species of squid, whereas Curtis and Cole used another. Hodgkin and Huxley used microelectrodes of one design, whereas Curtis and Cole used those of another.

Tolman's work on rat maze navigation illustrates the idiosyncrasy of data as well. Tolman (1948) reported the results of several "latent learning" experiments involving mazes of various designs. Blodgett (1929) had rats run a six-unit T-maze for a food reward, plotting error scores over multiple days of testing. Tolman and Honzik (1930) ran a fourteen-unit version of this experiment. Spence and Lippitt (1946) performed a Y-maze experiment in which rats chose between an arm with food and an arm with water.

Turn now from Bogen and Woodward's concept of data to their concept of phenomena. Woodward explains that,

> Phenomena, as I shall use the term, are relatively stable and general features of the world which are potential objects of explanation and prediction by general theory. Examples of real or putative phenomena, some of which will be discussed in more detail below, include weak neutral currents, gravitational radiation, Brownian motion, proton decay, capacity limitations and recency effects in short term memory, and the proportionately higher rate of technical innovation among middle-sized firms in moderately concentrated industries. (Woodward, 1989, p. 393)

One of the significant differences between data and phenomena is that, whereas data are supposed to be idiosyncratic to specific experimental contexts, phenomena are not. Phenomena are supposed to be "general features of the world."

I have not reviewed Bogen and Woodward on data and phenomena simply to take that framework on board. Instead, I use it to set the stage for

[5] See Curtis and Cole (1942).

articulating a slightly different picture. First of all, I propose to go beyond the idea that data are idiosyncratic to experimental contexts, such as Hodgkin and Huxley's lab at the Marine Biological Station at Plymouth. I propose, in addition, that data, as I intend them, are spatiotemporal particulars. They are individual measurements of, say, action potentials, that occur at specifiable times and places. To propose that data are spatiotemporal particulars is to go beyond proposing that they are idiosyncratic, since one might hold data are idiosyncratic types rather than tokens. For what it is worth, Bogen and Woodward might accept this further step, but that is inessential for my purposes. Here, I am specifying what I mean by data so as to provide a descriptively adequate account of the explanation of experimental results.

Second of all, I propose that for each spatiotemporal particular datum, a scientist aims to "hit" a spatiotemporal datum target or result. The idea is that the scientist aims to have one perceptually accessible spatiotemporal particular indicate something about another (typically) perceptually inaccessible spatiotemporal particular. More concretely, an oscilloscope datum might be intended to indicate, say, an action potential. Note that the use of "aims" is meant to allow that a datum may miss its target, as being confounded with other factors. One could put the matter another way: ideally, a single spatiotemporal localizable trace on an oscilloscope indicates a single spatiotemporally localizable action potential; ideally, a single verbal report indicates a single perception of the Hermann grid illusion. My concept of a result differs from what Woodward means by a phenomenon. Whereas my concept of a result is of a spatiotemporal particular, Woodward's concept of a phenomenon is something general.[6] This is how I understand Woodward's claim that phenomena are general features of the world. I do not reject his concept of a phenomenon. Instead, I am proposing another concept.

With the idea of a result on the table, it should be clear how the theory of singular compositional explanation dovetails with the concept of a result. In mid-twentieth-century physiology, some data were oscilloscope traces, whereas the results were action potentials. The results, in this case action potentials, were to be singularly compositionally explained in terms of ions fluxes. In the study of the Hermann grid illusion, the data were verbal reports, the results were perceptions and the contributing activity instances of those perceptions (recall Section 2.2.2) were to be singularly

[6] Paul Teller's concept of token phenomenon might be the same as my concept of a result (Teller, 2010, p. 820f).

compositionally explained in terms of the activity instances of retinal ganglion cells.

Further, the term "result" should be read as broad enough to include individuals, activity instances of individuals, rates of activity instances of individuals, and property instances of individuals. Much as Bogen and Woodward take their concept of phenomena to be ontologically noncommittal, so I take my concept of a result to be ontologically noncommittal.[7]

Before moving on, I need to digress to forestall a potential objection. One might think that Bogen and Woodward's view is that scientists do not explain data, as they are idiosyncratic, hence that scientists also do not explain results, as they too are idiosyncratic. At times, their writings might suggest this view: "Some versions of the claim that scientific theories explain facts about what we observe amount to the mistaken idea that theories explain facts about data" (Bogen & Woodward, 1988, p. 306) and "In Sections IV and V, we defend the claim that scientific theories explain facts about phenomena rather than data" (Bogen & Woodward, 1988, p. 314). There is, however, reason to think that they do admit that scientists *in some sense* explain data and that they would admit scientists *in some sense* explain results. This involves reading Bogen and Woodward's texts more carefully, reviewing the arguments they give for these views.

Consider, first, their case for saying that a theory of molecular structure does not explain a particular datum regarding the melting point of a sample of lead:

> It is easy to see that a theory of molecular structure which explains why the melting point of lead is approximately 327 degrees could not possibly explain why the actual data-points occurred. The outcome of any given application of a thermometer to a lead sample depends not only on the melting point of lead, but also on its purity, on the workings of the thermometer, on the way in which it was applied and read, on interactions between the initial temperature of the thermometer and that of the sample, and a variety of other background conditions. No single theory could accurately predict or explain an outcome which depends upon the confluence of so many variable and transient factors. (Bogen & Woodward, 1988, pp. 309–310)

The argument is not that *scientists* do not explain data. Nor is it that *science* does not explain data. Nor is it that there is no explanation of data at all. It is that no single theory explains a datum. To explain a datum requires resources beyond the scope of a single scientific theory. That, however, is

7 Bogen and Woodward (1988, p. 321).

consistent with there being *some sense* in which scientists explain data/results. It is consistent with there being *some sense* in which there are scientific explanations of data/results.

Second, the arguments Bogen and Woodward give in their Sections IV and V are somewhat different:

> Typically, scientific theories are expected to provide (what we shall call) systematic explanations of facts about phenomena rather than facts about data. To show why this is the case, we will briefly describe in this section two features which typical systematic explanations possess. The first has to do with the exhibition of detailed patterns of dependency. The second has to do with unification and systemization. (Bogen & Woodward, 1988, pp. 322–323)

Later, they claim, "it will often not be feasible to provide explanations of data satisfying the requirements on systematic explanation outlined above." Without going into what I take to be some of the subtle details of what systematic explanations are, the rejoinder is that when scientists explain data or results, they are not *systematically explaining* data or results. I can simply concede that explanations of data and results are not systematic. They are explanatory in another sense. Indeed, in a footnote in section IV, Bogen and Woodward open the door to such a possibility:

> In ordinary usage, the identification of a single factor which played a causal role in the production of some outcome is often described as an explanation of that outcome, even when many other factors also played a role in producing the outcome in question, and even when the details of the mechanism by which the identified factor produced the outcome are left vague and unspecified. Let us call such explanations "singular causal explanations." . . . Our point in introducing the requirements described below is not to challenge this usage. Rather our point is simply that such singular causal explanations differ in important ways from the kind of explanations we describe as "scientific" or "systematic," and which we intend our requirements to characterize. (Bogen & Woodward, 1988, p. 322, fn. 17)

Bogen and Woodward essentially invite the development of singular causal explanations of data. I have indirectly taken them up on this invitation, insofar as I have developed a theory of singular compositional explanation, not of data, but of results. Moreover, although I will not go into the matter, I can agree with them that singular compositional explanations differ in important ways from the explanations they describe as systematic.

Quite apart from Bogen and Woodward's view, there is good reason to be open to a theory of singular causal explanation and a theory of singular compositional explanation. I assume that scientists are not "mysterians" about the production of data or results. Although some data and results may be inexplicable, it is not generally true that scientists take the generation of data or results to be a complete and utter mystery inexplicable by science. Scientists generally believe that data and results are capable of explanation in some sense. My proposal is intended to flesh out the sense in which results are sometimes compositionally explained. Philosophers of science who are not mysterians about results should, therefore, be open to the possibility of singular causal explanation and singular compositional explanation.

3.3 Explaining the Results of Controlled Experiments

Consider now the typical context for the explanation of experimental results, namely, the explanation of the results of controlled experiments. By a "controlled experiment," I mean, in the simplest case, an experiment involving two sets of conditions that are taken to be the same for a single determinative factor and, possibly, a single determined factor.[8] Here, I will introduce the topic by way of the different explanation-seeking why-questions that controlled experiments can motivate.

Consider two explanation-seeking why-questions:

1. Why did axon no. 15 in seawater display an initial inward current when depolarized by 65 mV?
2. Why did axon no. 15 *in seawater* display an initial *inward* current when depolarized by 65 mV, but *in a sodium-free medium* display an initial *outward* current when depolarized by 65 mV?[9]

The first of these questions seeks a singular dynamic compositional explanation in which an activity instance of a whole (the inward current of axon no. 15) is explained in terms of activities of parts (the inward movements of sodium ions). The second question – call it a "controlled experiment question" – is obviously more complicated, as it seeks consideration of two distinct sets of conditions. An answer to this second question provides a singular dynamic compositional explanation of the inward current in the sodium-containing medium and a singular dynamic compositional explanation of an outward current in a sodium-free medium. In notation, there is an explanation of an instance of S Ψ-ing in terms of x_i's φ-ing and an

[8] My sense of "controlled experiments" is the same as that in Schickore (2019). These differ from what she describes as "controlling experiments," which involve experimentation meant to probe relevant variables.

[9] One might ask how this question relates to what are commonly called "contrastive questions." This issue is discussed in Chapter 9 in the context of Lipton's view of IBE.

explanation of an instance of S Ψ*-ing in terms of x_i's φ*-ing. Thus, the explanation of the results of a controlled experiment involves what I will loosely describe as an "amalgamation" of considerations beyond what is involved in a singular compositional abduction.[10]

As with any proposed philosophical distinction, there is room to debate exactly how to draw the distinction. Whatever one's theory it is important to notice that the compositional explanation of a result of a single experiment is not the same as the explanation of the results of a controlled experiment. Clearly a compositional explanation of S Ψ-ing is not identical to a compositional explanation of S Ψ-ing and S Ψ*-ing. Further, why-questions like 2) are raised by the results of controlled experiments used to test compositional hypotheses. The target of the distinction is the results of controlled experiments. To see this, return to the first experiment discussed in Hodgkin and Huxley (1952a). Chapter 1 broached the idea that Hodgkin and Huxley used an experiment to make an abductive argument supporting the sodium hypothesis. Now, however, I emphasize an additional detail of their experiment and their argument. In theory, Hodgkin and Huxley might have sought to use a singular dynamic compositional explanation – an answer to question "1" – in order to support the sodium hypothesis. They did not do that. Instead, as noted in Chapter 1, Hodgkin and Huxley performed an experiment in which they measured the current first in a seawater solution, then in a sodium-free solution, then again in a seawater solution. The results of this experiment invite questions like "2." To accurately describe their experiment, one must have something to say about the experimental control.

Hodgkin and Huxley's argument enabled them to convey to their fellow neuroscientists the crux of their thinking, namely, that the 65 mV depolarization somehow changes the membrane permeability. Thus, when the external medium has more sodium than does the axoplasm, sodium rushes in, whereas in a medium with less sodium than in the axoplasm, sodium rushes out. Hodgkin and Huxley's subsequent experiments regarding the sodium hypothesis varied the depolarization and the sodium concentration of the external medium, in essence, tracking how changes in the amount of depolarization and changes in the external sodium concentration affected the membrane current. The results of these experiments raise any number of more specific questions like "2."

Consider another example. Hodgkin and Huxley's second experiment in Hodgkin and Huxley (1952a) included a repetition of the first

[10] Thanks to Caitlin Mace for suggesting the term "amalgation" to me.

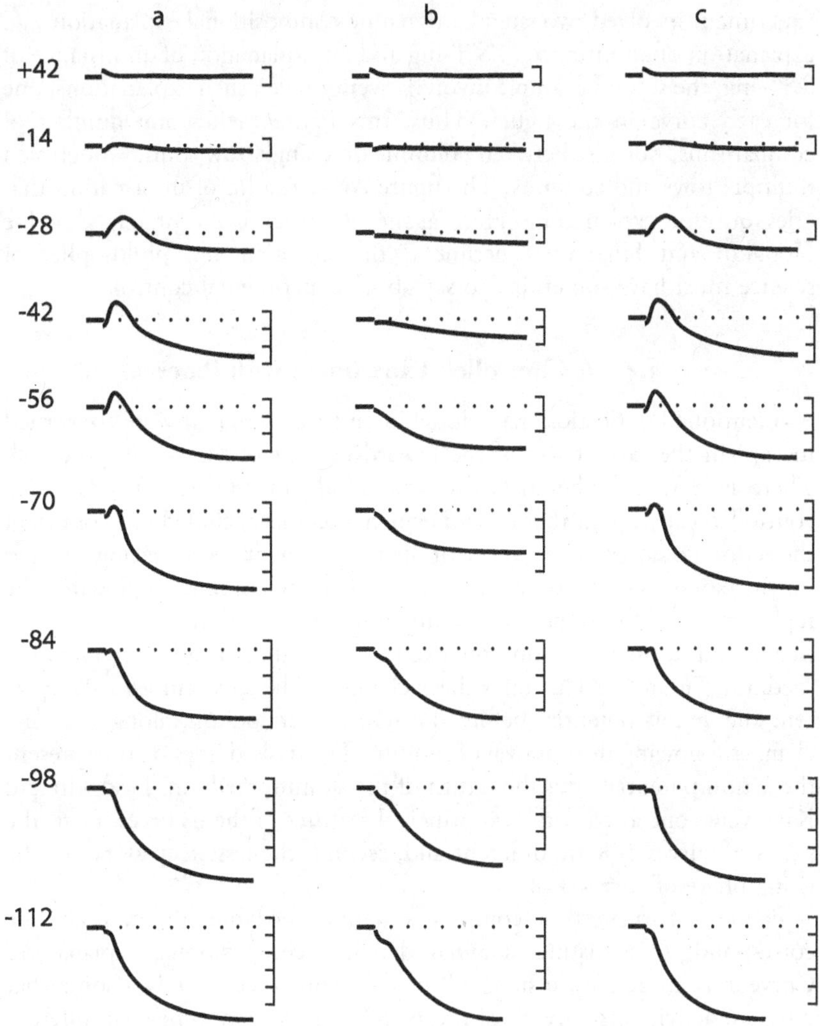

Figure 3.1 Ionic currents in response to different command voltages in sodium-containing (left column, a), sodium-free (middle column, b), and sodium-containing media (right column, c). Vertical scale: 1 division is 0.5 mA/cm^2. Horizontal scale: interval between dots is 1 msec. Redrawn from Hodgkin and Huxley (1952a, p. 451, figure 2).

experiment and its sodium-containing/sodium-free/sodium-containing protocol, only with the axon voltage-clamped at different values. The second experiment investigates the role of a second variable, the "command voltage" (see Figure 3.1). Whereas Hodgkin and Huxley's first experiment involved two singular dynamic compositional explanations, an explanation of an instance of S Ψ-ing and an explanation of an instance of S Ψ*-ing, the second example involves twenty-seven such explanations, one for each curve in the figure. Thus, this figure invites any number of comparisons, not just between columns of a single row, but also between multiple rows and columns. The figure conveys a lot of information that cries out for explanation. Here again, in order to accurately describe Hodgkin and Huxley's experiment, the historian and philosopher of science must have something to say about experimental control.

3.4 A Controlled Experiment with Rates

As mentioned in Section 3.1, Hodgkin and Katz (1949a) were concerned to explain the rate of rise of the inward current of the action potential. There is now a further detail to add: Hodgkin and Katz performed a controlled experiment that invited explananda that included rates of rise of the action potential. The essence of their experiment lies in tracing changes in an axon's activity over the course of a few minutes following the replacement of the standard sodium-containing medium with a sodium-free medium, then returning the axon to the standard sodium-containing medium. Figure 3.2 illustrates these changes. The solid curves 1–8 represent the action potential before the removal of sodium, along with the changes following the removal of sodium. The dashed lines 9–10 represent the action potential after the return of the sodium medium. Hodgkin and Katz were concerned with two principal features of these curves. First, the peaks of curves 1–8 are different and, second, the less steep slopes of the rising phase of curves 1–8.

Figure 3.2 suggests a complex situation. For one thing, each line corresponds to a distinct singular dynamic compositional explanation. Curve 1 is an activity instance Ψ of the axon S; curve 2 is a somewhat different activity instance Ψ* of the axon S; curve 3 is a somewhat different activity instance Ψ** of the axon S; and so forth. The singular dynamic compositional explanation of S Ψ-ing implicates one set of individual sodium ions moving into the axon, {x_1 φ-ing, x_2 φ-ing, . . ., x_n φ-ing}; the singular dynamic compositional explanation of S Ψ*-ing implicates a distinct set of individual sodium ions moving into the axon, {x_1 φ-ing, x_2 φ-ing, . . ., x_m φ-ing}; and the singular dynamic compositional explanation of S Ψ**-ing implicates yet another distinct set of individual sodium ions moving into the axon, {x_1 φ-ing, x_2 φ-ing, . . ., x_l φ-ing}. In the foregoing notation, the different letters in the final subscripts for the

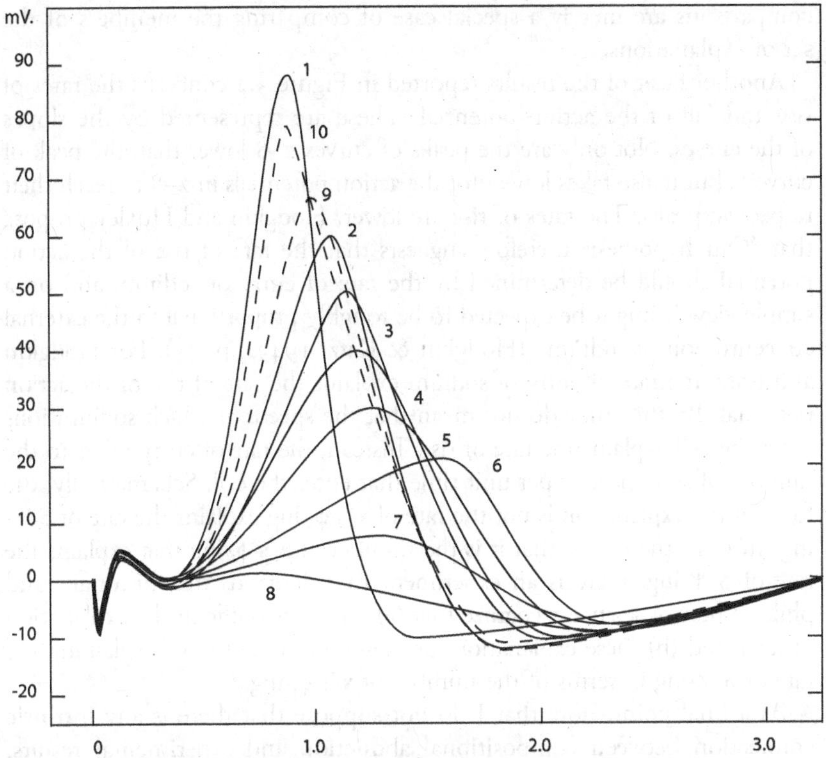

Figure 3.2. Action of isotonic dextrose. Record 1: action potential in sea water just before application of dextrose. 2–8: records taken at following times after arbitrary zero, defined as moment of application of dextrose: 2, 30 sec; 3, 46 sec; 4, 62 sec; 5, 86 sec; 6, 102 sec; 7, 107 sec; 8, 118 sec. Record 9 taken 30 sec after reapplication of sea water; 10, record at 90 and 500 sec after reapplication of sea water (only one curve is drawn since the responses at these times were almost identical).

individuals x are meant to indicate different numbers of ions moving across the membrane. (The notation does nothing, however, to capture the idea that two disjoint sets of ions might cross the membrane.) There is, thus, a set of singular dynamic compositional explanations. The figure clearly invites any number of pairwise comparisons, including the following: Why is the peak potential in curve 1 greater than the peak potential in curve 2? Why is the peak potential in curve 2 greater than the peak potential in curve 3? These are singular dynamic compositional explanations in a controlled experiment, much like that found in the first experiment of Hodgkin and Huxley (1952a). And, of course, the pairwise comparisons are merely a special case of comparing the members of the set of explanations.

Another facet of the results reported in Figure 3.2 concerns the rates of rise and fall of the action potential. These are represented by the slopes of the curves. Not only are the peaks of curves 2–8 lower than the peak of curve 1, but it also takes longer for the action potentials in 2–8 to reach their respective peaks. The rates of rise are lower. Hodgkin and Huxley propose that "Our hypothesis therefore suggests that the rate of rise of the action potential should be determined by the rate of entry of sodium, and on a simple view it might be expected to be roughly proportional to the external concentration of sodium" (Hodgkin & Katz, 1949a, p. 55). For Hodgkin and Katz, the rate of entry of sodium explains the rate of rise of the action potential. By this, they do not mean that the speeds at which sodium ions enter the cell explains the rate of rise. Instead, the rate of entry refers to the number of sodium ions per unit time that enter the cell. Schematically, the form of the explanation is not the rate of x_i's ϕ_i-ing explains the rate of S Ψ-ing. Instead, the idea is that it is the number of x_i's ϕ_i-ing that explains the rate of S Ψ-ing. Here is an experiment that demands that historians and philosophers of science recognize that (a) scientists traffic in the explanation of rates and (b) these explanations are sometimes matters of explaining the rate of S Ψ-ing in terms of the number of x_i's φ_i-ing.

As a final point, note that I do not suppose that there is any intrinsic connection between compositional abduction and experimental results. A scientist might use compositional abduction outside the context of experimental interpretation. For example, Chapter 6 will note that Baumgartner's introduction of the retinal ganglion cell theory of the Hermann grid illusion involved compositional abduction, but no new experimentation. Moreover, there can be experimental interpretations that do not involve compositional abduction. Causal abduction might play a role. Sometimes experimental interpretations might be limited to enumerative inductions. In a similar fashion, I do not suppose that there is any intrinsic connection between compositional abduction and the interpretation of the results of controlled experiments. Instead, it is merely that, in many cases, the scientific use of compositional abduction takes place in the context of the interpretation of the results of controlled experiments.

3.5 Summary

In this chapter, I have expanded on the compositional explanatory pluralism advanced in Chapter 2. I have proposed that scientists sometimes explain rates of activities of wholes in terms of the number of activities of associated parts. In addition, I have described cases in which scientists invoke compositional explanations of results, or data targets, that are produced in experiments. I have further proposed that there are explanations that arise from the results of controlled experiments. These explanations may be thought of as families of compositional explanations. I think it would be unsurprising to find that there are still other scientific explanations involving compositional relations, but further search for them would be orthogonal to my focus on compositional abduction. The expanded repertoire of explanations is merely the bare minimum that the historian and philosopher of science needs in order to describe the cases to be discussed in Chapters 5 and 6.

CHAPTER 4

Singular Compositional Abduction

Abductive reasoning infers some hypothesis based on what that hypothesis explains. Consider the following descriptions: abduction is "the operation of adopting an explanatory hypothesis" (Peirce, 1992, p. 231); abduction is "inference to an explanatory hypothesis" (Magnani, 2001, p. xxi); abduction is "reasoning from an observation to its possible explanations" (Aliseda, 2005, p. xii); "By 'abductive' inference I shall mean an inference where a central component of that inference is the fact that the inferred (purported) facts provide a putative explanation of the evidence or some part thereof" (Bird, 2005, p. 5). So understood, abductive inference differs from deductive inference in being defeasible. Further, abductive inference differs from simple enumerative induction insofar as the former, but not the latter, implicates explanation.[1]

I must emphasize that I really do understand abduction in the "thin" sense just described. Contrast this with two "thicker" senses that I do not adopt. First, there is one based on Peirce's writings. Peirce once proposed this schema for abduction:

> The surprising fact, C, is observed;
> But if A were true, C would be a matter of course,
> Hence, there is reason to suspect that A is true. (Peirce, 1992, p. 231)

Given that this is a schema for abduction, one might interpret Peirce to mean that the concept of abduction includes the concept of the explanandum C being surprising.[2] One might say that, for Peirce, it is analytic that,

[1] Cf., Douven (2017a).

[2] I do not mean to endorse this interpretation of Peirce. Here is why. Before introducing the schema, Peirce writes, "Long before I classed abduction as an inference it was recognized by logicians that the operation of adopting an explanatory hypothesis – which is just what abduction is – was subject to certain conditions" (Peirce, 1992, p. 231). So, maybe Peirce's concept of abduction is "thin" – abduction is adopting an explanatory hypothesis – but it is a further contingent empirical fact that abduction is only used to explain surprising facts. Under this interpretation, Peirce and I might share

in an abductive inference, the explanandum is surprising. So, on this interpretation of Peirce, an inference from some unsurprising explanandum would not be an instance of what Peirce would count as an abductive inference.

As I will illustrate on multiple occasions, there are instances where scientists draw inferences to compositional hypotheses based on what those hypotheses explain, but in which what is to be explained is not surprising to them. For these cases, and there are many of them, historians and philosophers of science need something other than the preceding interpretation of Peirce's concept of abduction. My thin concept of abduction fits the bill. With my concept of abduction, it is an empirical question whether, in any given instance, the explanandum is surprising. With my concept of abduction, I can show that, as a matter of empirical fact, there are cases in which the explanandum of an abductive inference is surprising, but other cases in which the explanandum is not surprising. Indeed, herein lies the rationale for using the thin concept: it facilitates an accurate description of many episodes of scientific reasoning.

My concept of abduction is also thin insofar as it does not assume that abduction is warranted abduction. On my thin concept of abduction, there can be abductive inferences that are warranted as well as abductive inferences that are not warranted. So framed, the distinction may be self-evident, but the distinction is not usually framed this way. Instead, in a move that has been widely overlooked in the subsequent literature, Harman (1965) proposed that if an inference to an explanatory hypothesis is to be warranted, then it must rule out rival hypotheses. He called such inferences to explanatory hypotheses that rule out rivals "inference to the best explanation." Thus, Harman's intent behind the concept of IBE is that it is a warranted inference – a warranted abductive inference.[3]

There are many instances where, on some page of a scientific journal article, scientists draw an inference to some compositional hypothesis based on what that hypothesis explains, but where it is not clear that, at that point in their published work, they take their explanatory hypothesis to be warranted. It could be that such warrant as scientists take there to be for some hypothesis emerges only through a prolonged program of scientific research, a program spanning more than just a few pages of a journal

a concept of abduction, but differ on how that reasoning is used in practice. Peirce would think it is only used to explain surprising observations, whereas I would not.

[3] Bird (2022) abandons the assumption that IBE provides knowledge, because of a concern that it cannot rule out defeating rival hypotheses. He, thereby, breaks with a founding assumption of the concept of IBE, although retaining many of the other views about it.

article. Scientists typically take a program of research, not one experimental result, to eliminate rival hypotheses. Here again, the rationale for using the thin concept is that it facilitates an accurate description of many episodes of scientific reasoning in the experimental literature.

Abductive inferences differ in terms of the explanations they invoke. Etiological abductions invoke etiological explanations. Analogously, singular compositional abductive inferences invoke singular compositional explanations. Thus, what is distinctive of singular compositional abductions, as opposed to, say, etiological abductions, is the reliance on singular compositional explanations as opposed to etiological explanations.

Section 4.1 develops the preceding picture. It describes some important features of abductive inference in general and illustrates how they apply to singular compositional abductive inferences more specifically. The applications sometimes include parts of the case studies from Chapters 5 and 6. The overarching conclusion of this section is that, to a first approximation, singular compositional abductive inferences are a species of the genus abductive inference.

Section 4.2 offers an account of why some scientists think that singular compositional abduction is truth-conducive. The gist of the scientific thinking is this. If a configuration of entities referred to in hypothesis H ontologically determines the configuration of entities referred to in some evidence E, then the entities of H must exist. Nonexistent entities do not make anything happen. But, if the entities of H exist – if they are as H represents them – then H is true. Given this conception, one can see why scientists take rival hypotheses to be such potent defeaters. The rival hypothesis is that it is not the configuration of entities referred to in H that determines the configuration of entities in E; rather, it is the configuration of entities referred to in H* that does this.

Section 4.3 contrasts how I understand abductive confirmation and HD confirmation. The idea, to a first approximation, is that just as historians and philosophers of science should abandon the DN model of explanation insofar as it is focused on logical relations among sentences rather than ontological relations among entities in the world, so they should abandon the HD model of confirmation insofar as it is focused on logical relations among sentences rather than ontological relations among entities in the world. When scientists explain, they implicitly presuppose that there are ontological dependence relations between what is explained and what is doing the explaining. When scientists abductively confirm, they implicitly presuppose that there are ontological dependence relations between what is confirmed and what provides the confirmation. In short, philosophers of science should make a more thoroughgoing break from the empiricist tradition.

Section 4.4 spotlights examples of abductive inferences in cognitive science wherein the explanans does not specify an individual bearing an activity or property instance. These examples differ in this regard from most of the inferences discussed in the case studies in Chapters 5 and 6. More importantly, the examples in this section will not be developed in fine detail by close and extended attention to the primary experimental literature. Consequently, they will play a different argumentative role in my larger project of examining how scientists confirm compositional hypotheses than do the examples from the case studies. Whereas the case studies are meant to support a detailed articulation and defense of a role for singular compositional abduction in science, the examples in this section are only meant to indicate areas of scientific research where the account of singular compositional abduction merits further attention. The examples in this section are evidence that singular compositional abduction is not some esoteric mode of scientific reasoning.

4.1 Four Features of Singular Compositional Abduction[4]

Singular compositional abductive inference in science, like abductive inference in general, has four features. (1) Scientists sometimes use singular compositional abduction in an effort to confirm some singular compositional hypothesis. (2) Scientists sometimes use singular compositional abductive inference to postulate entities that are qualitatively distinct from the entities cited in the supporting evidence. (3) The scientific use of singular compositional abductive inference may rely on background beliefs. (4) Scientists sometimes use singular compositional abductive inference to postulate entities that are not directly empirically detected. Although I take singular compositional abduction to be a type of abduction that has yet to be philosophically characterized, I propose that singular compositional abduction has these four features. Indeed, it is these commonalities that help place singular compositional abductive inferences alongside other abductive inferences.

1) *Abductive Confirmation.* In the abduction literature, there is some debate about the extent to which abduction provides confirmation. Some philosophers, such as Peirce, have proposed that abduction does not confirm a hypothesis but merely introduces some hypothesis as worthy

[4] The discussion of this section draws upon Aizawa and Headley (unpublished).

of further examination.[5] According to Peirce, other methods are required to confirm an abductively suggested hypothesis. Other philosophers, such as Douven, have proposed that abduction provides some degree of confirmation or support for the explanans hypothesis.

Recall from Chapter 1 that, strictly speaking, my concern is not with the extent to which abduction provides confirmation. I am not doing epistemology. My concern instead is with how scientists use abduction. On this score, there is good reason to believe that scientists, at least at times, take abductive inferences to provide confirmation.

There are three defeasible reasons to think that Hodgkin and Huxley assumed that their abductive reasoning surrounding the sodium hypothesis provides confirmation. First, they provided three compositional abductive arguments based on three sets of experimental results.[6] Assuming that the first argument sufficed to introduce the hypothesis, why would Hodgkin and Huxley provide two more arguments to do the same work? A confirmational interpretation of abduction, however, can at least offer a sketch of an answer. Each argument further supports the hypothesis; each argument provides further confirmation. How and why this is plausible merits further attention, but at least there is a start.

Second, after their first abductive argument, Hodgkin and Huxley did not stop giving abductive arguments, and then switch to some other form of argument to confirm the sodium hypothesis. The arguments consistently appealed to what the sodium hypothesis explains. On the Peircean idea that abduction merely introduces some hypothesis, it is unclear why Hodgkin and Huxley (1952a) present multiple abductive arguments involving the sodium hypothesis. On a Peircean view, one might expect Hodgkin and Huxley to adopt a different pattern of reasoning at later points in their 1952 papers. However, they did not. They continued to advance abductive inferences.

Third, after reviewing the principal results of their paper, Hodgkin and Huxley comment,

> These results support the view that depolarization leads to a rapid increase in permeability which allows sodium ions to move in either direction through the membrane. These movements carry the initial phase of ionic

[5] See, for example, Hanson (1960), Peirce (1992, p. 106), Hoffmann (1999), Niiniluoto (1999), Magnani (2001), Schurz (2008), Plutynski (2011), and Misak (2017). Some have proposed that while abduction does not provide confirmation, (warranted) abduction understood as inference to the best explanation does. See, for example, McAuliffe (2015), McKaughan (2008), and Mohammadian (2021). None of these sources explains why Peirce thought that scientists use abduction merely to introduce (pursuit-worthy) hypotheses. Nor do they consider whether scientists do in fact use abduction to introduce (pursuit-worthy) hypotheses.

[6] Hodgkin and Huxley (1952a, pp. 450–455).

> current, which may be inward or outward, according to the difference between the sodium concentration and the electrical potential of the inside and outside of the fibre. (Hodgkin & Huxley, 1952a, p. 471)

Suppose "support" means confirmation. If so, then Hodgkin and Huxley take the results used in the compositional abductive arguments to confirm the hypothesis that the depolarization leads to a rapid increase in membrane permeability and the initial inward phase of the ionic current.

Before moving on, consider a methodological aside. I have just proposed three defeasible reasons for thinking that Hodgkin and Huxley used abductive reasoning to confirm the sodium hypothesis. The first two reasons are based on features of the experiments that Hodgkin and Huxley performed and the argumentative structure of their paper. In contrast, the third reason appeals to Hodgkin and Huxley's assessment of what they had done. As one may recall from Chapter 1, my method is not simply to report what scientists say. Nor do I simply take scientific assessments at face value. So, I do not consistently accept what scientists say. Instead, I try to consistently adopt what, on balance, provides the best account of what the scientists are doing. I adopt what is supported by the preponderance of evidence. Sometimes I reject scientific meta-theoretic descriptions, as when scientists describe their work as instances of falsificationism or when scientists claim that science is not interested in explanation.[7] At other times, I accept scientific assessments. It is because of this "on balance" nature of the interpretation of scientific work that my discussion is sometimes more complicated than one might expect.

2) The Qualitative Distinctness of Explanans Entities and Explanandum Entities.

With an enumerative inductive inference, one might conclude that seawater has a conductivity of 50 mS/cm based on the observation that seawater sample 1 has conductivity of 50 mS/cm, seawater sample 2 has a conductivity of 50 mS/cm, and seawater sample 3 has a conductivity of 50 mS/cm. In this inference, there is no qualitative distinctness between the seawater in the conclusion and the seawater in the premises. Nor is there qualitative distinctness between the 50 mS/cm mentioned in the conclusion and the 50 mS/cm mentioned in the premises.

In contrast to enumerative induction, abduction allows for explanans entities that are qualitatively distinct from explanandum entities. Consider the textbook account of the explanation of marks in the snow. One might abductively infer that a deer walked through the area. A deer, of course, is

[7] Recall Raman and Ferster (2022, p. 85) and Pearson (1911, p. viii).

qualitatively distinct from marks in the snow. Consider two properly scientific examples. In the case of the singular compositional abductive arguments based on Hodgkin and Huxley's experiments, a current measured in axon no. 15 is qualitatively distinct from the movements of sodium and potassium across the cell membrane. An individual molecule of hydrogen fluoride has a property instance of a dipole moment; it is more negative on the fluoride side than on the hydrogen side. The molecule has this property instance in virtue of the property instances of the electronegativities of the hydrogen and the fluorine. Dipole moments are qualitatively distinct from electronegativity.

Another way to understand qualitative distinctness is by way of putative abductive inferences that *prima facie* do not respect this condition. Schurz describes "micro-part abductions" wherein "one abduces a hypothesis about the microscopic composition of observable objects in terms of micro-parts which obey the same laws as the observable macroscopic objects, in order to explain various observed empirical phenomena" (Schurz, 2008, p. 216). Although Schurz does not speak to qualitative distinctness, the sameness of laws might require the sameness of the property instances governed. So, for example, Coulomb's law differs from Newton's law of universal gravitation. Part of the difference between Coulomb's law and Newton's law of universal gravitation is that the first law covers charge, whereas the second covers mass. So, Schurz's microparts abductions *prima facie* violate the qualitative distinctness condition.

3) Abduction and Background Beliefs. Philosophers working on IBE have sometimes noted that background beliefs have a role to play in scientific IBE.[8] New Mechanists have also commented that background beliefs play an important role in science.[9] The account of singular compositional explanation developed in Chapter 3, however, enables us to articulate in more detail how background beliefs relate to abductive inferences. What follows will not be a full accounting of the role of background beliefs in scientific reasoning, but something closer to a set of preliminary observations that relate background beliefs and singular compositional abduction.

For concreteness, let us consider, yet again, the Hodgkin–Huxley singular compositional explanation of one instance of the initial inward current of axon no. 15. The explanandum of this explanation is the inward

[8] Cf., for example, Day and Kincaid (1994), Niiniluoto (1999), Salmon (2001), Lipton (2003), Schurz (2008), Norton (2021), and Davey (2023).

[9] See, for example, Craver and Darden (2013, pp. 8, 11, 47, 64, 69, 81, 85, 107, 115, 124) and Craver et al. (2021, p. 8088).

current of the neuron; the explanans includes the influx of sodium ions. Let us say that the set of foreground beliefs of this singular compositional explanation are the beliefs that refer to the individuals, their activity instances, the ontological dependence relation(s) between them, and the background conditions (such as temperature and pressure) that appear in the explanans and explanandum.[10] In other words, the foreground beliefs of a singular compositional explanation are beliefs about the individuals and activity instances that, under the given background conditions, stand in the compositional ontological dependence relation of implementation in a singular compositional explanation.

The foreground beliefs in a singular compositional abduction are, naturally enough, the foreground beliefs in the singular compositional explanation implicated in the abductive inference. It should be noted that, on the assumption that beliefs are representations, the term "foreground beliefs" refers to the representations involved in an explanation. These observations connect this chapter's theory of foreground and background beliefs from the abduction literature to Chapter 2's theory of compositional explanations as representations of compositional ontological dependence relations from the philosophy of explanation literature.

The Hodgkin–Huxley example involves an explanation of an activity instance in terms of other activity instances. The specification of foreground beliefs should be more general than this. It should not be limited to the representations in singular dynamic compositional explanations; it should include at least the representations in singular analytic compositional explanations and singular standing compositional explanations. Recall Table 0.1. Thus, I propose this generalization: the foreground beliefs of a singular compositional explanation are the beliefs about the individuals, property instances, activity instances, rates of activity instances, ontological dependence relations, and background conditions that occur in a singular compositional explanation. Further, the foreground beliefs of a singular compositional abduction are the foreground beliefs of the implicated singular compositional explanation.

The background beliefs for a given abduction are all the "non-foreground" beliefs a given scientist may hold. This is a deliberately broad categorization. Clearly, background beliefs for a given abductive inference are relative to the given inference. Change the abductive inference and one will change the beliefs that are in the background. Given the specification

[10] To avoid confusion, background *beliefs* are part of a scientist's psychology, whereas background *conditions* are parts of the world a scientist studies.

of background beliefs, it should be unsurprising that different background beliefs stand in different relations to the foreground beliefs invoked in a given abductive inference. Given the open-ended specification of background beliefs, it is likely that background beliefs figure in a lot of scientific reasoning. Thus, to rephrase a point I made above, I do not aspire to a complete taxonomy of all scientific reasoning with background beliefs. Instead, I will give examples of background beliefs impinging on the explananda entities, explanantia entities, background conditions, and relation between explananda and explanantia entities involved in an abductive inference. To a first approximation, background beliefs are involved at every turn.

Recall from Chapter 3 the distinction between data and results. In Hodgkin and Huxley's experiment on axon no. 15, the data are what appeared on the oscilloscope, whereas the results are the currents in the axon. One explanandum of Hodgkin and Huxley's singular compositional abduction using axon no. 15 is the initial inward current in the axon. The belief that there is an initial inward current is a foreground belief. In earlier experimental work, Hodgkin and Huxley made the case that what registered on their oscilloscope – the data – provided tolerably reliable evidence about the axonal currents – the results.[11] In other words, Hodgkin and Huxley's earlier investigations supported their background belief that the oscilloscope registers the currents in axons. This background belief supported the explanandum belief that, during a single spatiotemporally localizable voltage clamping, axon no. 15 had an initial inward current. This is a case in which a background belief supports an explanandum belief of an abductive inference.

In the experiment with axon no. 15, the beliefs about the sodium concentrations in the external media were foreground beliefs. For the seawater conditions, Hodgkin and Huxley added local seawater to the apparatus holding the axon. Thus, Hodgkin and Huxley's foreground beliefs about the extracellular sodium concentration in the seawater conditions of the experiment were informed by a background belief about the concentration of sodium in the seawater around Plymouth. This background belief supported the part of the explanans having to do with the sodium concentration of the medium surrounding axon no. 15 during one voltage clamp. By contrast, Hodgkin and Huxley's foreground belief about the extracellular sodium concentration in the sodium-free condition were informed by a background belief about how they had prepared their

[11] See, for example, Hodgkin and Huxley (1945) and Hodgkin et al. (1949).

substitute "choline seawater."[12] In this case, the supporting background belief was not a general hypothesis about the concentration of sodium in the local seawater, but a belief about the process by which they mixed the choline seawater. Distinctly different background beliefs supported the different foreground beliefs about the sodium concentrations of the external media.

The distinction between data and results applies again with regard to the background condition of temperature. A pervasive background belief in physiology of the day was that thermometer readings (data) reliably track temperatures (results). Hodgkin and Huxley do not bother to try to document the reliability of their thermometer in the way they tried to document the reliability of their oscilloscope and microelectrodes. Instead, they merely note their foreground belief that the experiment with axon no. 15 took place at 11° C. Here is a case in which a background belief supports a hypothesis about background conditions.

Hodgkin and Huxley had the foreground belief that sodium ions moving across the membrane of axon no. 15 would generate a current in the neuron. It was supported by an instantiation of a background belief about what electrical currents are. Current just is the movement of charge. In the case at hand, this was a belief about how activity instances of many individuals at one level implement an activity instance of another individual at another level. Here, there is a case where a background belief supports a hypothesis about a compositional ontological dependence relation between explanans entities and explanandum entities.

In the last few paragraphs, I have noted how background beliefs may support foreground beliefs in a singular compositional explanation, hence how background beliefs may be said to support a singular compositional abduction. All this said, the vast majority of a scientist's beliefs are likely to be inferentially irrelevant to a given explanation. So, *prima facie*, the beliefs that the earth is about 93 million miles from the sun and that the human population of the earth in 1900 exceeded one million are likely to be inferentially irrelevant to the explanation of the initial inward current of the action potential.

4) Abduction and the Unmeasured. It is sometimes proposed that, with abduction, one might postulate entities, not based on direct measurement or detection of them, but because they are explanatory. Sometimes this point is made by historical examples drawn from archeology, geology, or evolutionary biology, wherein one does not have direct access to events in the distant past. A scientist might, for example, explain the "iridium

[12] See Hodgkin and Huxley (1952a, p. 450) for their description of the ionic concentrations of this solution.

anomaly" – an abundance of iridium in a rock layer at the Cretaceous–Paleogene boundary – by appealing to a meteor impact. Because a meteor impact would explain the iridium anomaly, some scientists think they have reason to believe in a meteor impact.[13] Lipton (2014, p. 227) makes the point in terms of galaxies receding being an explanation of the red shift of light from those galaxies.

Singular compositional abduction has this feature as well. The Hodgkin-Huxley abductive inferences appeal to ion fluxes because they are explanatory. In the Hodgkin-Huxley case, the basis for the inaccessibility is not that the explanans events are in the distant past, but that the ions are exceptionally small and their activities are exceptionally fast.[14] Given the technology of their day, Hodgkin and Huxley could measure the total current across the axonal membrane, but they could not track the movements of individual sodium or potassium ions. Nor could they even directly track the movements of individual ion species. The best they could do was infer that, during action potentials, certain ions were moving according to their concentration gradients and reversal potentials, based on what such movements would explain.

The earlier comments on background beliefs and abduction may suggest that philosophers of science need a more nuanced description of the way in which a singular compositional abduction goes beyond what is directly measured. If one abductively infers from the iridium anomaly that there was a meteor impact, one likely has background beliefs that there are meteors, they sometimes strike the earth, they eject debris that settles over much of the earth's surface, and so forth. What is unmeasured in this inference is the role of a meteor in the explanation of this individual layer of sediment. In the inference to the hypothesis of galactic recession based on the red shift of light, galaxies and movements are not absolutely unmeasured entitles. What is unmeasured in this abductive inference are the red shifts of specific galaxies. In the Hodgkin–Huxley example, what was abductively inferred was the specific role of sodium ion fluxes in the initial inward current of the action potential.

4.1.1 *Other Examples of the Scientific Use of Singular Compositional Abduction have these Four Features*

The Hodgkin–Huxley example shows a case of scientific singular compositional abduction with the four features introduced above. Now I will

[13] Bird (2022, p. 173) uses this example as an illustration of IBE.

[14] Lipton might accept this case as an instance of what he calls "vertical inferences" (Lipton, 2003, pp. 66–68; Lipton, 2014, p. 227).

provide two more examples of compositional abductive inference that share these features: two of Tolman's experiments on rats' latent learning of mazes and Baumgartner's retinal ganglion cell theory of the Hermann grid illusion.

Tolman and Latent Learning. Tolman (1948) provides an interpretation of a series of experiments with rats navigating a diversity of mazes. A common view at that time was that rats navigated mazes by responding to local features of the mazes. The rats relied on a repertoire of stimulus–response pairs. Tolman, however, proposed that rats navigated using cognitive maps of the mazes that they built up through their interactions with the mazes. According to Tolman,

> [T]he central office itself is far more like a map control room than it is like an old-fashioned telephone exchange. The stimuli, which are allowed in, are not connected by just simple one-to-one switches to the outgoing responses. Rather, the incoming impulses are usually worked over and elaborated in the central control room into a tentative, cognitive-like map of the environment. And it is this tentative map, indicating routes and paths and environmental relationships, which finally determines what responses, if any, the animal will finally release. (Tolman, 1948, p. 192)

Two "latent learning" experiments—one a replication of the other—illustrate how Tolman believed cognitive maps were responsible for the rats' navigational abilities. The first version of the experiment used the six-unit T-maze show in Figure 4.1. Group I was a control group in which the rats were released in the start box and allowed to find their way to the food. Figure 4.2 shows that these rats quickly learned to run the maze with few errors. Group II rats were released into the start box and allowed to find their way around the maze but without food being placed in the food box for the first six days. On the seventh day and on each subsequent day, however, the rats were given food in the food box. The "x" above the dashed line in Figure 4.2 indicates when the rats received food. Finally, Group III rats were released into the start box and allowed to wander through the maze without food being placed in the food box. On the third day, however, they were provided with food. This point is indicated by the "x" above the dotted line in Figure 4.2. Tolman interprets his results as follows:

> It will be observed that the experimental groups as long as they were not finding food did not appear to learn much. (Their error curves did not drop.) But on the days immediately succeeding their first finding of the food their error curves did drop astoundingly. It appeared, in short, that during the non-rewarded trials these animals had been learning much more

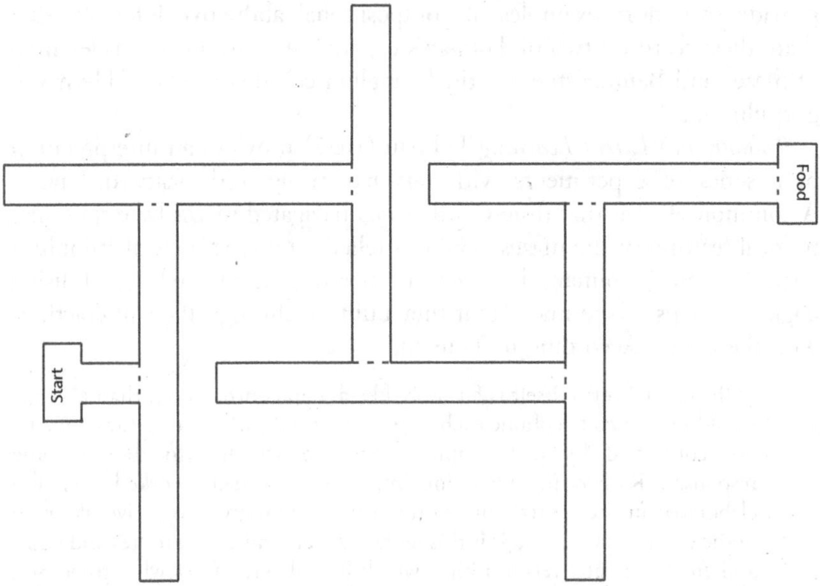

Figure 4.1 Six-Unit Alley T-Maze, redrawn from Tolman (1948, p. 193, figure 4).

> than they had exhibited. This learning, which did not manifest itself until after the food had been introduced, Blodgett called "latent learning." (Tolman, 1948, p. 194)

In the replication, Tolman replaced the six-unit maze in Figure 4.1 with a larger fourteen-unit maze. He also gave the rats more days to learn. The control group, again, received food throughout. One test group never received food, whereas a second test group received food on the eleventh day onwards. Tolman was not surprised to find that, after the eleventh day, the errors in the second test group declined dramatically.

These experiments illustrate the four features of abductive inference that I have described. To begin with, Tolman takes these experiments to support the cognitive map hypothesis, rather than merely introducing it. Again, as with the Hodgkin–Huxley example, there are three reasons in support of this conclusion. First, the six-arm T-maze would serve to introduce the cognitive map hypothesis, so it would be unnecessary to run the fourteen-unit replication study to introduce, yet again, some hypothesis that had already been introduced. Second, Tolman does not use singular compositional abduction with the six-arm T-maze to introduce the cognitive map hypothesis, but then switch to some other method

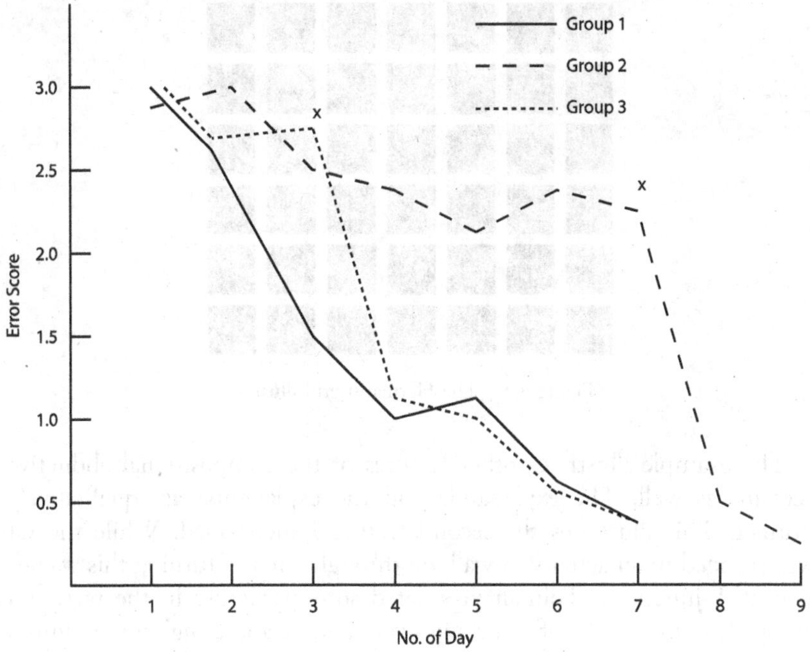

Figure 4.2 From H. C. Blodgett, The effect of the introduction of reward upon the maze performance of rats. *Univ. Calif. Publ. Psychol.*, 1929, 4, No. 8, p. 120. Redrawn from (Tolman, 1948, p. 193, Figure 5).

to confirm the hypothesis. He uses the same pattern of reasoning for both mazes. Third, and finally, in some of the prefatory remarks regarding the experiments, Tolman comments: "The [experiments], out of many, which I have selected to report are simply ones which seem especially important in reinforcing the theoretical position I have been presenting" (Tolman, 1948, p. 193). If "reinforcing" is similar in meaning to confirming, then Tolman interprets his experiments to offer confirmation of the cognitive map hypothesis.

Recall the methodological clarification that I registered above in connection with the Hodgkin–Huxley example. Two of the reasons I just offered were based on features of the experiments, most notably that the fourteen-unit T-maze experiment replicates the six-unit T-maze experiment. The third, however, was based on Tolman's self-assessment. I treat Tolman's interpretation of what he is doing as a defeasible reason for his thinking that he is offering confirmation. His treating abduction as confirmatory makes sense of some of the arguments (Tolman, 1948).

Figure 4.3 The Hermann grid illusion.

The example illustrates other features of the compositional abductive account as well. The explananda and the explanantia are qualitatively distinct. This illustrates the second feature I mentioned. While the rat was engaged in an activity of walking through a maze, turning this way or that at T-junctions, Tolman postulated some structure in the rat brain engaged in an activity of "working over" and "elaborating" inputs into a cognitive-like map. The rat behavior was qualitatively distinct from the "working over" and "elaborating."

Third, the design and interpretation of one experiment was shaped by background beliefs. Although Tolman does not mention it, the original experiments were controlled for both odor and tactile cues, so that rats could not rely on them to find food.[15] Again, this background knowledge enables Tolman to eliminate the rival hypotheses that the rats were navigating, not using a cognitive map, but by local olfactory or tactile cues. Fourth, and finally, Tolman could not directly detect the relevant brain structure or its activity. All he manipulated or measured in the experiments was the rat or its behavior. The cognitive map was entirely hypothetical; a hypothesis justified by its being explanatory.

Baumgartner's Retinal Ganglion Cell Theory of the Hermann Grid Illusion. In 1870, Ludimar Hermann published a brief description of an optical illusion bearing his name[16] (see Figure 4.3). In one version of the Hermann grid illusion, there are intersecting vertical and horizontal white bars on a black field wherein scintillating gray smudges appear at the intersections somewhat peripheral to the viewer's fixation point. (In

[15] Blodgett (1929, p. 118). [16] Hermann (1870).

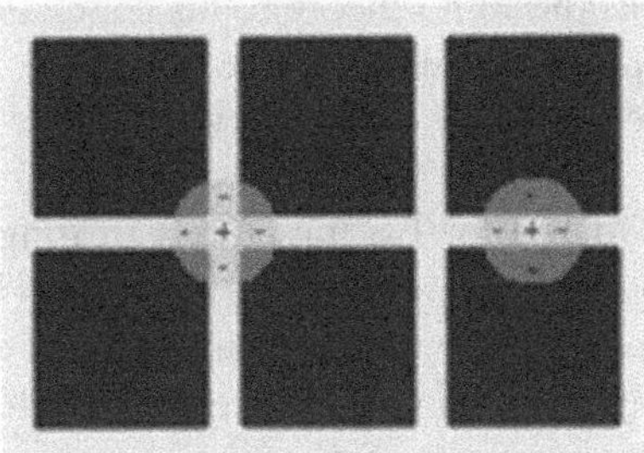

Figure 4.4 The RGC explanation of the Hermann grid illusion.

another version, there are black bars on a white field wherein scintillating gray smudges appear.) During the 1950s, physiologists recording the response properties of cells in the optic nerve discovered retinal ganglion cells (RGCs) that were activated by lights shown in their central regions but suppressed by lights shown in a surrounding annulus.[17] These came to be described as ON-center/OFF-surround cells. In a paper in 1960, Baumgartner proposed that these cells can be part of an explanation of the illusion. The idea is that when an ON-center/OFF surround cell falls at an intersection, there is more OFF-surround stimulation, thereby making for a weaker cell response and a "smudge" (see the left half of Figure 4.4). By contrast, when an ON-center/OFF-surround cell falls along a line away from an intersection, there is less OFF-surround stimulation, thereby making for a stronger cell response and no "smudge" (see the right half of Figure 4.4). Baumgartner apparently believed that RGCs form the biological basis of the Hermann grid illusion since the activities of these cells would explain the perception of the illusion.

This example has the four features of abductive inferences I have described. To begin with, Baumgartner does not merely take the RGC theory to be worthy of further investigation. He apparently assumed that it is correct, so that he can immediately use the RGC theory as a basis for investigating another question, namely, the size of receptive fields. He concludes his very brief paper writing, "The size of the foveal receptive fields and the size of the receptive fields of each other retinal area can be calculated from the width and distance of the object. In humans, this determination method in the foveal area yields a diameter of the receptive fields of >25μ" (Baumgartner, 1960, p. 22). Baumgartner does not

[17] Cf., for example, Kuffler (1953).

articulate how he links the RGC theory to his conclusions about receptive field sizes, but the point remains that he takes the RGC theory to be sufficiently warranted that it may serve as the basis for further conclusions.[18]

Notice that the evidence for Baumgartner's treating abduction as confirmatory is unlike what was found in the Hodgkin–Huxley and Tolman cases. For these former cases, there were variations on an experiment. The Baumgartner case does not have this. Further, in the Hodgkin–Huxley and Tolman cases, there were reports of what the scientists themselves thought they had done. The Baumgartner case does not have this either. Instead, Baumgartner just assumed that his abductive conclusion about the neuroscientific basis of the Hermann grid illusion is correct, then proceeded to reason further based on that assumption. He does not try to provide further confirmation of the RGC theory.

Now consider qualitative distinctness. The perception of the Hermann grid is qualitatively distinct from the RGCs. The grid is perceived to have scintillating gray smudges; the RGCs have no perception. The perception of the grid involves horizontal and vertical white bars; the RGCs have no perception. The RGCs have radially symmetric receptive fields; the perception does not. To put matters in another way, in offering a compositional abduction in support of the RGC theory, one appeals to entities and activities not found in the explanans.

It bears emphasis that Baumgartner's conclusion was not based on any experimental work that he reported in the paper. This provides an especially compelling example of the influence of background beliefs and the inessential role of empirical access to the lower level entities at work in producing the higher level entities. Baumgartner did not present a Hermann grid to a subject. Nor did he in any way stimulate a neuron or measure the activity of a neuron. Whether or not Baumgartner had the technical ability to perform an experiment in which he measured the activity of a participant experiencing the grid illusion, he, in fact, did not do such an experiment.

Instead of performing a new experiment, Baumgartner simply relied on background beliefs based on the results of prior experimental work. Although Baumgartner provides no references to the scientific literature, enough is known about the relevant history of science to enable us to determine the basic structure of the relevant background work. On the one

[18] Spillmann (1971) provides a more detailed account of determining receptive field sizes on the assumption of the RGC theory.

hand, there was a tradition of creating visual stimuli that had some of the contrast characteristics of the Hermann grid illusion and then showing them to participants and, in one way or another, measuring their responses. On the other hand, there was a physiological tradition of probing the receptive fields of neurons by projecting small spots of light onto the retina. A simple way of making the point is that Baumgartner drew his conclusion based exclusively on his familiarity with a lot of background information.

4.2 Why Some Scientists Think Singular Compositional Abduction is Truth-Conducive

Tolman, Hodgkin, Huxley, and Baumgartner, among others, thought that their compositional abductive reasoning provided grounds for believing that the world was configured in some way. More specifically, they thought that their abductive reasoning provided grounds for believing that an activity instance of one individual was implemented by activity instances of some other lower level individual or individuals. I now want to spell out how I think they intended to connect singular compositional abduction with truth. Just to be clear, I am not providing an account of why singular compositional abduction is truth-conducive. I am doing a form of history and philosophy of science, not epistemology. Instead, I am providing an account of why scientists thought they had a truth-conducive inference method.

The proposal begins with the contention that the scientists in the cases at hand never embraced the empiricist philosophy of explanation. Maybe some scientists have embraced explanatory empiricism, but none of those in my case studies have. Informally speaking, for these scientists, explanation was not a matter of displaying logical relations among true sentences. These scientists instead embraced a realist philosophy of explanation. For these scientists, compositional explanation was showing how certain things in the world bring about certain other things. More technically, the idea is that these scientists treated compositional explanation as revealing compositional ontological dependence relations among entities in the world.

Explanatory realism paves the way for historians and philosophers of science to see how scientists link singular compositional explanations to confirmation and truth. If a scientist explains a particular spatiotemporally localizable axon current in terms of particular spatiotemporally localized ion fluxes, the scientist presupposes the existence of those ion fluxes. If the ion fluxes did not exist, then they would not be what explains the axonal current.

The explanation presupposes that the existence of certain entities does determinative work. It is this presupposition that invites the step from H explains E to the truth of H. If what is mentioned in H is what makes what is mentioned in E occur, then what is mentioned in H had better exist.[19]

Given this account, one can also see why scientists would treat "failure" to explain as disconfirming. Begin with an example. In 1939, Hodgkin and Huxley used internal electrodes to measure the magnitude of the action potential. Julius Bernstein's "membrane theory" explained that action potentials were generated by a breakdown in membrane permeability that allowed ions to flow down their concentration gradients. This theory has it that the action potential should approach zero. More technically, on this theory, the breakdown of the membrane will ontologically determine that the action potential approaches zero. What Hodgkin and Huxley found was that the action potential overshot zero. The membrane theory ontologically determined something that did not, in fact, materialize. This served as disconfirmation of Bernstein's membrane theory, in time paving the way for the confirmation of Hodgkin and Huxley's theory of the action potential.

On the realist account of the scientific rationale underlying abductive inference, one can also see why scientists treat rival hypotheses as such potent defeaters of abductive inferences and why scientists have such frequent recourse to them. (Here, I assume that philosophers of science familiar with experimental scientific literature will readily acknowledge this feature of scientific practice. My case studies in Chapters 5 and 6, however, will document the importance of rival hypotheses.) The idea is this. Given that, at some point in time, both H and H' explain E, scientists often assume they have no reason to prefer H to H' or H' to H. Either the entities referred to in H or the entities referred to in H' might just as well determine E, so there is no reason to prefer H to H', and *vice versa.* Chapters 5 and 6 show that one scientific strategy for overcoming this temporary impasse is to conduct further experiments that will enable a scientist to show that H explains E, E', E", . . ., whereas H' fails to explain, say, E".

Here, I have proposed that scientists take the "bare" explanation of some singular explanandum to have confirmation-theoretic import. Scientists suppose that the ontological determination of some entities by other entities has confirmation-theoretic significance. One might say that explanatory considerations *narrowly construed* have confirmation-theoretic import. This narrow view stands in contrast to a view that maintains that simplicity, whatever that is, is an explanatory virtue and that it has

[19] Cf., Cartwright (1982, pp. 14–16).

confirmation-theoretic import. One version of this view is that the simplicity of an explanation, whatever that is, exhausts the confirmation-theoretic import of an explanation. Another version is that the simplicity of an explanation, whatever that is, provides additional confirmation-theoretic import to an explanation. On either version, one might expect that if scientists think that the simplicity of an explanation of a singular explanandum has confirmation-theoretic significance, then they would draw attention to that simplicity. To put the point concretely, if Hodgkin and Huxley thought that the simplicity of the explanation of the sodium hypothesis had some confirmation-theoretic import, then they would have said something about it. They would have drawn attention to the simplicity of the explanation when making the case for the sodium hypothesis. Yet, Hodgkin and Huxley did not mention the simplicity of the sodium hypothesis explanation. Why? In this case, they did not believe that the simplicity of the explanation had confirmation-theoretic import. The point here is not to deny that simplicity ever has confirmation-theoretic significance. It is instead to draw attention to cases where there is reason to think it does not. This is not a totally obvious conclusion, but nevertheless a reasonable interpretation of what Hodgkin and Huxley were up to. And what has been said about simplicity might also be said about some other putative virtues of single explanations, such as what is ad hoc, plausible, and so forth.[20]

4.3 Abductive Confirmation and HD Confirmation

Realism about scientific explanation and abduction provides for a theory of confirmation that constitutes an advance over HD theories of confirmation.[21] For one thing, an abductive theory of confirmation can accurately describe many cases of scientific reasoning that are *prima facie* instances of HD confirmation. Further, and perhaps more importantly, an abductive theory of confirmation can avoid the "tacking problems" that have beset theories of HD confirmation for decades. Indeed, an abductive approach offers a diagnosis of the source of the tacking problems: the logical relations in standard first-order logic with identity do not capture what it is for some bit of evidence to be relevant to some hypothesis. The relevance

[20] Scope is a somewhat different case, since the scope of some hypothesis has to with its role in multiple explanations, rather than just one.

[21] Lipton (2003) was concerned to argue that his version of IBE was superior to HD confirmation.

relation is not logical; it is ontological. The relevance relation is one of ontological dependence.

To flesh this out, let us reconsider Carl Hempel's classic introduction to the philosophy of science which introduced the now well-known example of Semmelweis's discovery of the cause of childbed fever.[22] In describing the case, Hempel mentions how one or another hypothesis did or did not explain some observation.[23] He then interpreted this scientific practice in terms of his explanatory empiricism. Hempel (1966) did not discuss the Semmelweis case in terms of the DN model, but his philosophy of science book did appear the year after his monumental anthology on scientific explanation.[24] Hempel proposed that (some?) scientific disconfirmation is the application of *modus tollens*:

If H is true, then so is [E]
2a] But (as the evidence shows) [E] is not true.
H is not true. (Hempel, 1966, p. 31)

(Some?) scientific confirmation is the application of the fallacy of affirming the consequent:

If H is true, then so is [E].
2b] (As the evidence shows) [E] is true.
H is true. (Hempel, 1966, p. 31)

In later years, the HD account of confirmation assumed a familiar, more expansive form. Evidence E confirms hypothesis H given background knowledge K if and only if

i) H.K is consistent
ii) H.K entails E
iii) K alone does not entail E.[25]

As with DN explanation, HD confirmation proposes to dispense with ontological relations among entities in the world in favor of logical entailment relations among true sentences.

A realist version of scientific explanation and abduction provides an alternative account of why scientists sometimes engage in a type of hypothetical reasoning. The hypothetical reasoning, on its own, does not

[22] Hempel (1966). Hempel's example presupposes a causal explanation, but, as I will show, the points carry over to compositional explanation.
[23] Hempel (1966, p. 29). [24] Hempel (1965).
[25] For some prominent discussions, see Glymour (1980a), Gemes (1993), and Sprenger (2011).

provide confirmation. Instead, the hypothetical reasoning draws attention to ontological dependence relations – relations empiricists eschew. Sometimes this ontological dependence relation obtains between hypothesized explanans entities and empirically detected explanandum entities. In very simple cases, the ontological dependence implicit in the argument is evident to scientists and goes without independent motivation. In more complicated cases, the dependence between what is referred to in the hypothesis H and what is referred to in the consequent E is not evident. In such cases, it requires substantial scientific reasoning to establish the connection.

Here is another way of illustrating the divergence between the empiricist and the realist interpretation of Semmelweis's hypothetical reasoning. When rejecting the atmospheric-cosmic-terrestrial explanation of puerperal fever in the two clinics he supervised, Semmelweis made the following comments:

> But if the atmospheric-cosmic-terrestrial conditions of Vienna cause puerperal fever in predisposed persons, how is it that for many years these conditions have affected persons in the first clinic while sparing similarly disposed persons in the second? (Semmelweis, 1983, p. 65)

> If [influences] operate outside the hospital, certainly those who are admitted to the first clinic will be no more subject to them than those admitted to the second. No significant difference in mortality between the equally exposed patients admitted to the two clinics would then exist. (Semmelweis, 1983, p. 66)

> On the other hand, if epidemic influences operate on individuals during their stay in the hospital, there could be no difference in the mortality rate, since two clinics so near one another that they share a common anteroom must necessarily be subject to the same atmospheric-cosmic-terrestrial influences. (Semmelweis, 1983, p. 66)

> If the atmospheric influences of Vienna occasion an epidemic in the maternity hospital, then necessarily there must be an epidemic among maternity patients throughout Vienna because the entire population is subject to the same influences. But in fact, while the puerperal disease rages most furiously in the maternity hospital it is only infrequently observed either in Vienna at large or in the surrounding countryside. (Semmelweis, 1983, p. 66)

In reading these passages, empiricists adopting the HD confirmation assume that historians and philosophers of science can abstract away for Semmelweis's use of "cause," "operate," and "occasion" as real relations among things in the world. What matters for empiricists, like Hempel, is a logical relation among true sentences. In contrast, the realist proposal is that the terms "cause," "operate," and "occasion" refer to an important ontological element in Semmelweis's reasoning, namely, an ontological dependence relation between things in the world. At later points in his discussion, Semmelweis does not always use such terms, but he does use them often enough to provide the historian and philosopher of science some reason to interpret at least a significant portion of his reasoning as abductive.

There is a scientific practice – a type of scientific hypothetical reasoning that lends itself to interpretation by historians and philosophers of science as HD confirmation. This reasoning should not, I propose, be interpreted in terms of HD confirmation, but in terms of abductive confirmation. Let me now provide a concrete illustration that is closer to the mid-twentieth-century physiological work that will be the subject of Chapter 5. This example, from Curtis and Cole (1940), differs from Hempel's puerperal fever example by invoking compositional, rather than causal, explanation, but is otherwise quite similar. Moreover, the example recommends itself as about as manifestly an instance of HD confirmation as one might wish from a real scientific example.

Curtis and Cole assume that action potentials move the same way from electrode **a** to electrode **c** as from electrode **c** to electrode **a**:

> If, as we are assuming, the action potential difference recorded between the needle and an outside electrode opposite its tip is due entirely to the membrane action potential, it should be independent of the direction of propagation of the impulse. To test this, the axon was stimulated first at an **a** and then at the **c** electrode, and both action potentials recorded on the same film. In several cases the potentials superimposed so completely as to be inseparable . . . It is quite apparent that the amplitudes and forms of the action potentials are nearly identical for propagation in either direction. (Curtis & Cole, 1940, pp. 151–152, bold added for clarity)

This brief passage clearly involves hypothetical reasoning. The hypothesis is that the action potential difference recorded between the needle and an outside electrode opposite its tip is due entirely to the membrane action potential. From this hypothesis, Curtis and Cole "deduce" a potential "observation," namely, that the potential difference should be independent of the direction of the propagation of the impulse. In this example, I assume that the conditional was evident to Curtis and Cole, and their readers, so that they felt no need to justify it. Curtis and Cole, then, performed an experiment in which they stimulated at **a** then at **c**, finding that the action potentials are "nearly identical for propagation in either direction."

The theory of HD confirmation takes a step beyond merely reporting what Curtis and Cole did. The theory of HD confirmation offers a philosophical "gloss," theory, or interpretation of what Curtis and Cole did. The gloss, theory, or interpretation is that Curtis and Cole confirmed the hypothesis by displaying a relation among true sentences – they provide an instance of affirming the consequent – along the following lines:

1. If the action potential difference recorded between the needle and an outside electrode opposite its tip is due entirely to the membrane action potential, then the action potential is independent of the direction of propagation of the impulse.
2. The action potential is independent of the direction of propagation of the impulse.[26]

   Therefore, the action potential difference recorded between the needle and an outside electrode opposite its tip is due entirely to the membrane action potential.

A realist abductive account of this scientific practice, however, proposes that the theory of HD confirmation omits a crucial part of what underlies Curtis and Cole's reasoning. It omits a role for ontological determination relations. A realist interpretation proposes that Curtis and Cole were not focused on sentences; they were focused on things in the world, namely, action potentials and their propagation. Realist abduction proposes that Curtis and Cole's phrase "is due entirely to" is their terminology for "is ontologically determined entirely by." Thus, the worldly hypothesis Curtis and Cole entertain is that the measured potentials are (entirely) ontologically determined by the membrane action potential. The worldly consequence of that hypothesis is that action potential results should be the same whether the stimulus is at **a** or at **c**. To further flesh out the abductive account, one might say that the experimental results of stimulating at **a** and at **c** raise the explanatory why-question, "Why is the action potential produced with stimulation at **a** the same as the action potential produced with stimulation at **c**?" The answer is that both results are produced by the same process.[27]

[26] Notice that what is to be explained here is why the action potential is independent of the direction of propagation. This explanandum is not an individual, an activity instance of an individual, or a property instance of an individual. Thus, it is unlike any of the explanations described in Chapter 3. Nor is it like any of the explanations in Chapter 4. Here the methodology of engaging closely with actual scientific reasoning supports greater explanatory pluralism as well as greater abductive pluralism.

[27] Note that Curtis and Cole must be assuming an equivalence between the action potential in the **a** stimulation case and the action potential in the **c** stimulation case. Strictly speaking, the action potential in the one case is numerically distinct from the action potential in the other.

The realist view of explanation and abduction embraces the idea that scientific practice includes hypothetical reasoning and that this might be interpreted as HD confirmation. In the face of that, the realist view maintains what looks like HD confirmation is really something else. This may be the thought underlying these comments by Clark Glymour:

> [T]he hypothetico-deductive account remains today one of the most popular. The reason, I think, is that it is so obviously the correct account of a great deal of the history of science. There can scarcely be a doubt that people have repeatedly claimed to provide evidence for some theory by deducing it from a testable proposition that turned out to be true, and that such arguments have not seemed odd or incorrect to their contemporaries. How this can be so is an important question for any alternative account of confirmation, but nonetheless the proper conclusion from the failure of hypothetico-deductive accounts of evidential relevance is that we must somehow locate a different structural connection between evidence and theory. (Glymour, 1980a, pp. 47–48)[28]

I agree with Glymour that scientists, like Curtis and Cole, have often provided evidence for some theory by (loosely speaking) deducing true consequences of the theory and that fellow scientists have abided such claims. There is such a practice. Where Glymour and I part ways is in Glymour thinks that these practices involve bootstrap confirmation, whereas I propose that they sometimes involve abductive confirmation.

The abductive approach I have developed provides a descriptively adequate account of certain episodes of scientific reasoning. I have provided one neat illustration of this from Curtis and Cole, but I will provide more in Chapters 5 and 6. Another advantage of the abductive account of confirmation, however, is that it offers a diagnosis of the "tacking problems" facing HD accounts of confirmation.[29] Let me review this now.

First, there is the problem that HD confirmation apparently allows tacking on irrelevant conjuncts in the premises, thereby HD-confirming irrelevant conjuncts. Let X be any arbitrary sentence, such that H.K.X is consistent. If H is HD confirmed by E, then so is H.X. Follow this through the schema. E confirms a hypothesis H.X given background knowledge K if and only

i) H.K.X is consistent
ii) H.K.X entails E
iii) K.X alone does not entail E.

[28] See, also, Lipton (2003, p. 15).

[29] For a clear, compact recent discussion, see Sprenger (2011).

Second, there is the problem of tacking on irrelevant disjuncts in the conclusion. If E HD confirms H, then so does E ∨ E'. E ∨ E' confirms a hypothesis H given background knowledge K if and only if.

i) H.K is consistent
ii) H.K entails E ∨ E'
iii) K alone does not entail E ∨ E'

This is because any proof of E could be extended in one step to a proof of E ∨ E' by disjunction introduction.

An abductive approach to confirmation can provide a diagnosis of the tacking problems: what makes a conjunct in the premises or a disjunct in the conclusion irrelevant is that the entities referred to in the conjunct/disjunct are not among the relata of the ontological dependence relation. Logical entailment relations in first-order logic with identity do not respect ontological dependence relations among things in the physical world. (Recall from Chapter 2 the role of ontological dependence in offering a diagnosis of some of the famous counterexamples to the DN model of explanation.) Take the problem of irrelevant conjuncts. In cases where worldly objects represented by K.H ontologically determine E, one cannot assume that something else in the world arbitrarily represented by some X will also ontologically determine E. The Hermann grid illusion, E, may ontologically depend on retinal ganglion cells (H), but not on retinal ganglion cells and the tide in Hudson Bay (H.X). The tide in Hudson Bay is ontologically irrelevant to the perception of the Hermann grid illusion. With the problem of irrelevant disjuncts, retinal ganglion cells may ontologically determine the Hermann grid illusion, but not the disjunction of the Herman grid illusion or the precession of the perihelion of Mercury. There is a logical entailment from a sentence describing the Hermann grid illusion to a sentence describing the Hermann grid illusion or the precession of the perihelion of Mercury, but this should not be confused with the existence of an ontological dependence relation between entities in the world.

Before moving on, there may be some expository value in contrasting my realist theory of explanation and abduction and an HPS-inspired "science in practice" theory recently advocated by Yafeng Shan. Shan (2020) proposes that "the hypothetico-deductive (H-D) model is no longer the mainstream account of evidence in the philosophy of science. Nevertheless, in the history (and even contemporary practice) of science, the H-D model has been ... widely used. So there is a gap problem between the philosophical analysis of evidence and good actual scientific practice" (Shan, 2020, p. 159).[30] On my realist view, this "gap problem" is resolved by claiming that scientists are not using the HD model of confirmation. Instead, scientists are engaged in hypothetical reasoning, but this reasoning is only part of the story. This hypothetical reasoning reveals ontological dependence relations between explanans entities and explanandum entities. Empiricists, and pragmatists such as Shan, ignore or deny this scientific commitment to ontological determination relations.

Shan proposes to solve the tacking problems, among others, with a revised version of HD-confirmation. Shan's proposal is the following:

> An H-D evidential practice p is a good practice if p fulfills the following conditions in order to show e confirms h.
>
> (I) The logical condition: In p, the evidential proposition e is a logical consequence of the hypothesis under test h and the background information k (i.e. $h \wedge k \models e$), while the background information k alone does not entail the evidential proposition e (i.e. $k \nvDash e$);
>
> (II) The practical condition: p is a series of repeatable scientific activities (e.g. conceptualisation, experimentation, and reasoning), which reliably and accurately justifies the evidential proposition e.
>
> (III) The contextual condition: In p, the interpretation of the evidential proposition e and background information k are acceptable and all of e, k, and h are relevant within its historical context. (Shan, 2020, p. 167)

The logical condition (1) corresponds to familiar formulations of HD-confirmation, whereas (II) and (III) add elements from philosophy-of-science-in-practice. Condition (II) is a requirement that the experimental practice p justify the proposition e. Condition (III) is a requirement that the practice specify that e, k, and h are relevant. For present purposes, the idea is that scientific practice informs philosophers of science that e, k, and h are relevant.

Here, I do not intend a critique of Shan's view, as this would take me far afield. Instead, my goal is to use his account as an expository foil. Shan's account is not wrong so much as it is incomplete. Here, I focus on (I) and (III). By my lights, the logical condition retains the core problem of hypothetico-deductive confirmation: it omits the role of ontological dependence relations. (I) provides only part of an empirically adequate description of scientific practice. Why, one might ask, would scientists

[30] Betz (2013, p. 991) has an earlier version of the gap problem.

think logical relations among true sentences tell them about how things happen in the world? How would logical relations among true sentences give scientists confirmation about how things happen in the world? The abductive realist proposes that, on their own, they do not. Instead, the logical relations at best draw one's attention to ontological dependence relations among entities in the world. Such ontological dependence relations back explanations and explanations, in turn, back abductive inferences.

The contextual condition (III) also leaves out an important element of an accurate description of scientific practice. Grant that history reveals what is, or is not, confirmationally relevant. There still remains the question, "What is this relevance relation that historical context reveals?" What is historical context telling us about? History may tell us that some fact is irrelevant, but it does not tell us what this irrelevance is. To put the matter more concretely, history may tell us that scientists think the influx of sodium ions into the axon is relevant to the initial inward current of the action potential, but it does not tell us what this relevance is. Shan does not have an account. By contrast, realism proposes that the relevant entities stand in an ontological dependence relation to the initial inward current. This ontological dependence relation has the various features described in Chapter 2, namely, they are many-one, asymmetric, transitive, contemporaneous, mass-energy neutral, natural dependence relations among entities that are not wholly distinct.

The short moral of this section is that historians and philosophers of science should not limit their philosophical interpretation of scientific hypothetical reasoning to logical relations among sentences. Ontological dependence relations have a part to play in some of these philosophical interpretations. One should not, however, read a stronger view into these comments. The view is not that scientific hypothetical reasoning just is compositional abductive reasoning; the view does not espouse a "philosophical reduction" of scientific hypothetical reasoning to compositional abductive reasoning. For one obvious thing, some hypothetical reasoning likely involves causal abduction. For another nonobvious thing, there are instances of hypothetical reasoning that involve compositional relations, but not by way of the compositional abductive inferences described so far. As these instances are not obvious, they will receive their own description in Chapter 10. In short, the discussion in this section will surely not be the last word on the relationship between HD confirmation and ontological dependence relations, but it does broach a proposal that is worth more serious consideration than it has been given.

4.4 Singular Compositional Abduction in Cognitive Science

Singular compositional abductive inferences are based upon singular compositional explanations. In these explanations, a scientist explains, say, an activity instance Ψ of some individual S in terms of the activity instances φ_i of some lower level parts x_i. There are, however, a great many abductive inferences in cognitive science in which the explanations are not so complete. I have in mind explanations in which scientists do not identify the lower level individual or individuals x_i in the explanans. These explanations have "individual-free" explanantia.[31] Importantly, abductive inferences involving explanations that have "individual-free" explanans are quite common in cognitive science. Here, I will provide a sample.

Tolman's explanation of rat maze navigation in terms of a cognitive map has an individual-free explanans. Tolman believed that rats navigate at least some mazes, under certain circumstances, using cognitive maps of the maze. He did not, however, specify a brain structure or structures, x_i, that functioned as a cognitive map. The first serious attempt to localize Tolman's cognitive maps came decades later, quite by accident, with John O'Keefe's discovery of what came to be called "place cells" in the hippocampus. Tolman was not, however, the only psychologist to abductively interpret the behavior of rats running mazes.

Consider now, not just a single explanation, but an entire field of cognitive science. There is a tradition of generative linguistics in which it is common to present native speakers with strings of words to elicit their judgments about them. These look to be intralevel experiments in which one intervenes on a native speaker, and then measures some response. Such experiments are often used in an attempt to learn something about the structure of a sentence, which is in turn thought to illuminate something about the structure of a putative language faculty. Noam Chomsky provides a simple illustration of how one might run a *bona fide* experiment along these lines:

> If a sentence as such as "flying planes can be dangerous" is presented in an appropriately structured context, the listener will interpret it immediately in a unique way, and will fail to detect the ambiguity. In fact, he may reject the second interpretation, when this is pointed out to him, as forced or unnatural (independently of which interpretation he originally selected under contextual pressure). Nevertheless, his intuitive knowledge of the language is clearly such that both of the interpretations (corresponding to

[31] Perhaps these are a special case of what New Mechanists sometimes call "mechanism sketches."

> "flying places are dangerous" and "flying planes is dangerous") are assigned to the sentence by the grammar he has internalized in some form. (Chomsky, 2014, p. 21)

Here, there is an analogy between Tolman's experiments with rats running mazes and humans interpreting sentences. Chomsky postulates certain cognitive structures because those structures would explain the behavior of native speakers. On the basis of such examples, Chomsky postulated that native speakers possess a grammar that generates two distinct parse trees corresponding to the two interpretations of the presented sequence of words. Nevertheless, Chomsky cites no individual in the brain that computes the two parse trees corresponding to the string "flying planes can be dangerous." He invokes an individual-free explanation.

Consider Miller's famous "Magic Number Seven" as applied to the span of human memory.

> In the simplest test of mnemonic capacity, a sequence of symbols (usually decimal digits) is read aloud or shown to the person at a regular rate (usually one per second) and at the end of the sequence he is asked to repeat or write the symbols in the correct order. The experimenter begins with short sequences and increases the length until the person is no longer able to repeat the entire sequence without error. This point is called the "span of immediate memory." (Miller, 1956, p. 131)

Here is an experiment from which Miller wishes to draw some conclusion about the structure of human short term memory. These experiments, run on humans, could not involve invasive manipulations or measures on the human brain. Moreover, Miller did not try to identify a brain structure implementing short term memory. He invoked an individual-free explanation.

Finally, consider an example from more recent cognitive science. Susan Carey postulates that the adult human mind has two systems for object individuation.[32] The first is a fully conceptual system that individuates objects in terms of kinds, such as dogs, cats, and cups. The second system individuates objects in terms of spatiotemporal information, that is, how the objects move through space. Show a participant Panel A, then five minutes later show the participant Panel B (see Figure 4.5). Carey proposes that the participant would judge that the square has moved from the lower left to the upper right, whereas the star has moved from the upper left to the lower right. By contrast, have participants fixate on the large black spot in the middle of A and B, then rapidly alternate between A and

[32] See Carey (2009, pp. 72–73).

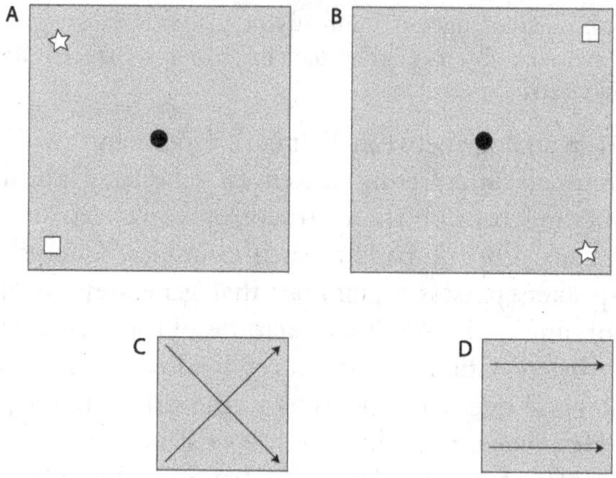

Figure 4.5 Two competing bases of solving the problem of numerical identity. The problem is which individual in Panel A is the same individual as which individual in Panel B. If numerical identity is traced relative to kind or property, one sees a star and a square moving diagonally (Panel C). If numerical identity is determined by minimizing the total amount of motion, one sees a square turning into a star or vice-versa (Panel D). Redrawn from Carey (2009, p. 73, figure 3.2).

B. In this situation, one perceives two objects moving horizontally across the stage. Carey's interpretation of these proposed intralevel experiments is that the conceptual kind-based system is active in the first situation, whereas the spatiotemporal system is active in the second. Again, there is an analogy between the Tolman and Hodgkin–Huxley cases, on the one hand, and the Carey case, on the other. Not the least of these is that Carey does not suggest a brain structure that implements the two types of object representation.

Our last cases based on Chomsky, Miller, and Carey are not close readings of scientific interpretations of actual experimental results. They are not like the Tolman and Hodgkin–Huxley cases. They are only meant to be suggestive. The suggestion is that there is a lot more science that can be understood in terms of abduction, wherein scientists postulate unmeasured, undetected things because those things explain. This is a proposal that merits attention.

4.5 The Singular versus the General

I have proposed that there are singular scientific compositional explanations, explanations of one spatiotemporal particular in terms of other spatiotemporal particulars. To return to my standard example, there were many initial inward currents of Hodgkin and Huxley's axon no. 15 measured one summer day in 1939. To a first approximation, at least, Hodgkin and Huxley assumed that each of these currents was to be explained in terms of an influx of sodium ions. These inferences are from the singular to the singular; they are from the particular to the particular. I further proposed that these explanations are used in singular compositional abductive inferences.

Of course, not all scientific inferences are from the singular to the singular.[33] There are inferences from the singular to the general. From a sample of experiments on axon no. 15, Hodgkin and Huxley, of course, went on to generalize. They may have drawn many distinct generalizations. They might have generalized, by a simple enumerative induction, from the idea that *some* initial inward currents of axon no. 15 were due to an influx of sodium to the conclusion that *all* initial inward currents of axon no. 15 were due to an influx of sodium. Maybe they also generalized to the conclusion that all initial inward currents of squid axons were due to an influx of sodium ions. Maybe they generalized even more broadly to a conclusion that all initial inward currents of axons were due to an influx of sodium ions. In many cases, statistical methods will guide the generalizations. This Hodgkin and Huxley generalization is based on a sample from a single individual, but there are also clear cases of inferences from a sample drawn from multiple distinct individuals. In all of these cases, historians and philosophers of science need to distinguish the singular compositional abductive inferences from "neighboring" inferences that they may interact with.

In the foregoing, I hedged my claims about generalization by claiming that Hodgkin and Huxley might have done this or might have done that. The hedge merely reflects my intention to be noncommittal about this, because those inferences *prima facie* go beyond the abductive inferences that are the focus of this book. One conjecture, not explored here, is that the generalizations – the various general compositional hypotheses – are inferred by enumerative induction from the singular compositional hypotheses. *Prima facie* how scientists justify general compositional hypotheses falls outside the scope of an account of singular compositional abduction. The qualification *prima facie* arises because I wish to sidestep Harman's thesis that warranted enumerative inductive is a special case of inference to the best explanation.

[33] Edouard Machery suggested the following paper as just the tip of the iceberg illustrating some of the diversity of reasoning outside the scope of the singular-to-singular inferences found in singular compositional abduction: Fisher et al. (2018).

Just to emphasize the limitations on the scope of my investigation, I might add that not all scientific inferences are from the singular to the singular or from singular to the general. Sometimes scientific inferences are from generalizations to particulars. A scientist may use the general compositional hypotheses that the initial inward currents of action potentials are implemented by sodium influxes to infer the singular compositional hypothesis that the initial inward current of *this* action potential is implemented by an influx of sodium. Here again, such inferences fall outside the scope of a theory of singular compositional abduction.

4.6 Summary

Section 4.1 reviewed material that should be familiar to philosophers who have studied abduction. It described features that singular compositional abduction shares with other instances of abduction. Scientists sometimes use singular compositional abduction to confirm the existence of hypothetical entities that are not directly measured and that are qualitatively distinct from evidential entities. Such inferences may be informed in multiple ways by background beliefs.

Sections 4.2–4.4 developed new ideas. Section 4.2 focused on the work that compositional ontological determination relations play in abductive reasoning. For scientists, ontological determination links an abductive inference to truth. If the entities referred to in H are to ontologically determine the entities referred to in E, then the entities in H had better exist. The hypothesis H had better be true. The account provides a rationale for why scientists (1) accept abductive inferences as truth-conducive, (2) believe that failure to explain disconfirms, (3) treat rival hypotheses as potent defeaters of hypotheses, and (4) do not often appeal to the simplicity of an explanation when offering an abductive argument. A lot of scientific practice, thus, makes sense on the abductive approach.

Section 4.3 provides a philosophical interpretation of scientific hypothetical reasoning. Such reasoning sometimes reveals ontological determination relations. Section 4.3 also made the case that an appeal to ontological determination relations addresses at least some of the challenges facing more familiar HD accounts of confirmation.

Section 4.4 draws attention to yet another type of abductive inference, namely, abductive inference that relies on explanations that have an "individual-free" explanans. In these cases, scientists will likely wish to know what individuals belong in the explanans. They will wish to be in a better epistemic state. They do not, however, reveal scientists thinking that the individual-free explanations are in some sense deficient, defective, or diminished.[34]

Finally, Section 4.5 emphasized that the theory – to this point – has been limited to singular inferences. These are inferences from one singular fact to another. A complete theory of scientific inductive inference would *prima facie* need to add another theory for inductive generalizations and for inferences about particulars from generalizations.

[34] Compare this conclusion with Craver and Kaplan (2020).

CHAPTER 5

The Compositional Basis of Resting and Action Potentials

From the opening paragraph of this book, I have referred to Hodgkin and Huxley's reasoning about an experiment to illustrate a set of scientific interpretations that I have been striving to characterize. In this chapter, I locate that experiment, and some of the others I have discussed, in the context of a case study of the role of singular compositional abduction in the development of the Hodgkin–Huxley theory.

Section 5.1 notes that Bernstein's membrane theory represented the received view of the resting and action potentials for much of the first half of the twentieth century. According to this theory, excess intracellular potassium ions generated the resting potential of a neuron, whereas a breakdown of the axonal membrane allowing ions to flow out of the cell generated the action potential. In 1939, Hodgkin and Huxley (HH) used intracellular electrodes to measure an "overshoot" of the action potential, indicating to them that the action potential could not be a mere dissipation of the potassium concentration gradient. This is an instance of abductive disconfirmation. It also reviews the work by Hodgkin, Huxley, Curtis, and Cole (HHC&C) to confirm that the overshoot was a *bona fide* experimental result. This subsequent work illustrates the importance of the data/results distinction from Chapter 3.

Section 5.2 describes how, having satisfied themselves that the overshoot was real, HHC&C looked more closely at quantitative measures of the effects of potassium on axonal function. Here, I pay special attention to HH's work on potassium in Hodgkin and Huxley (1946, 1947), since it very nicely illustrates controlled experiments in which the results are compositional abductively interpreted.

Section 5.3 reviews two key developments in HH's new theory. By 1947, HH had formulated an alternative to Bernstein's membrane theory: a rapid influx of sodium generates the rising phase of the action potential, whereas a delayed efflux of potassium generates the falling phase and the potential reversal. One key development in Hodgkin and Katz (1949a) was the replication of the "Overton experiments," which established a role for sodium in the action potential (see Section 5.3.1). Another key was a series of experiments using "direct methods" of radioactively labeling sodium and potassium to verify their movements during axonal activity.[1] This work provides background information that informed HH's interpretation of the electrophysiological experiments. Further, a philosophical theory of this work needs to recognize a role for the explanation of rates, in this case, rates of radioactive decay and rates of ionic migration (see Section 5.3.2).

Section 5.4 reviews a selection of experiments from HH's 1952 papers. As this discussion focuses on compositional abduction, I have made two significant choices of material to exclude. First, I review the experiments on sodium, but ignore those on potassium. The principal reason for this is that, for technical and theoretical reasons, the determinations of potassium currents involved much more hypothetical reasoning that is inessential to the theory of compositional abductive inference. Second, I exclude from this review any significant discussion of the mathematical Hodgkin–Huxley model from the fifth paper in the series. Although a good case can be made for thinking that this is the most scientifically important work of the series, it is not where singular compositional abduction looms large. Instead, it is concerned with formulating a mathematical model, a topic involving any number of philosophical issues beyond abduction.

Much of the story arc for this chapter will be unsurprising to those who are familiar with the history of mid-twentieth-century physiology and the Hodgkin–Huxley papers of 1952. The novelty in the story is the philosophical account of singular compositional abduction. The story will spell out how the account developed in Chapters 2–4 articulates much of what is implicit in HHC&C's reasoning about axonal activity.

5.1 Bernstein's Membrane Theory and Its Disconfirmation

In 1902, Bernstein proposed that cells exhibit potentials due to selective ion permeability. This membrane theory was predicated on earlier work by Walther Nernst, who showed that electrical potentials could be established by ionic concentration gradients, and the theory of Wilhelm Ostwald, who showed that the voltage across a membrane arises from its selective permeability to specific ions.[2]

[1] Keynes (1948, 1949), Keynes and Lewis (1951a).

[2] Nernst (1889), Ostwald (1890). For further history, see Lenoir (1986), Seyfarth (2006), and De Palma and Pareti (2011).

The Nernst equation relates a potential across a membrane to the relative concentration of a particular ion, denoted *ion*, inside and outside the membrane.

$$E_{ion} = \frac{RT}{zF} \ln \frac{[ion]_i}{[ion]_o}.$$

In this equation, E is the potential across the membrane for a particular ionic species (e.g., K^+ or Na^+), R is the gas constant, T is temperature in Kelvin, z is the charge of the ion, and F is Faraday's constant. The square brackets signify the ionic concentration with an o subscript for the outside (extracellular) concentration and an i subscript for the inside (intracellular) concentration.[3]

Inserting known values into the equation, Bernstein showed that the Nernst equation captures the relationship between the values on the right of the equation and the membrane potential. Bernstein further reasoned that, if the membrane potential is, in fact, determined in the way indicated by the Nernst equation, then one should find that the membrane potential varies with temperature. And, by measuring the potential at temperatures ranging from -2°C to 36°C, Bernstein confirmed that this relation held.[4]

Consider how this example fits the theory of singular compositional abduction sketched in Chapter 4. An individual membrane potential is a property instance of a neuron. It is ontologically determined by the property instances of the charges on the potassium ions and their relative internal and external concentrations under the conditions specified by R, T, z, and F. For each temperature, Bernstein stimulated the nerve generating a measurable "injury current," which he measured using a "differential rheotome."[5] Here, the injury current results provided evidence of a further result, the membrane potential. For each membrane potential result, there is a distinct set of potassium ions that realizes that membrane potential. Further, taken collectively, the set of estimated potentials for the temperatures from -2°C to 36°C constitutes a controlled experiment. Finally, Bernstein's hypothetical reasoning – if the membrane potential is determined in the way suggested by the Nernst equation, then one should find that the membrane potential varies with temperature – enabled him to provide support for the hypothesis that membrane potentials are determined in the way described by the Nernst equation. Part of the story of this support is compositional abduction.

[3] Contemporary forms of the Nernst equation have the outside concentration as the numerator and the inside concentration as the denominator. The presentation here follows Bernstein's convention.

[4] See Seyfarth (2006, pp. 5–6). [5] Seyfarth (2006, pp. 4–5).

Turn now from the resting potential to the action potential. Bernstein proposed that the action potential was simply a breakdown in the membrane that dramatically increased permeability to ions. Theoretically, this would diminish the electrical potential across the membrane, driving it toward zero. Membrane resistance or intracellular anions might prevent the potential from falling entirely to zero, but it should approach zero.[6] Thus, on the membrane theory, the magnitude of the action potential should be roughly equal to that of the resting potential.

Bernstein's membrane theory remained the standard view of resting and action potentials until January 1939. At that time, the 24-year-old Hodgkin had acquired new experimental equipment and "thought it best to begin with a straightforward problem, partly to run-in the new equipment and partly because [he] wanted to leave time for other activities" (Hodgkin, 1994, p. 130). Hodgkin sought to confirm the membrane theory in readily available shore crabs *Carcinus maenas*. Using external electrodes to measure the relative values of resting and action potentials, he found that the latter were significantly larger than the former.[7] If this result were correct, then it indicated that Bernstein's membrane theory was mistaken. He quickly appreciated the significance of the unexpected result:

> [A]s I wrote to my mother on February 6 the result was quite different: 'My experiments have taken on an interesting but rather disconcerting turn. I started off to work out a problem and thought beforehand that one particular result was a foregone conclusion. Now I have found the exact opposite. Of course I may discover a mistake or complication of some kind, but if not, this result means a fairly drastic reorganization of current ideas about nerve'. (Hodgkin, 1994, p. 131)

This autobiographical comment is helpful for two reasons. First, it confirms that Hodgkin quickly saw the significance of the result. By the summer of 1939, HH had devised a system for measuring action potentials in squid giant axons with intracellular electrodes.[8] With this system, they confirmed the overshoot. In their initial brief publication of the result – less than two full columns in *Nature* – HH did not indicate the theoretical significance. However, the autobiographical comment reveals that Hodgkin quickly understood the ramifications. Second, by the time HH could perform the intracellular recording, neither was likely to be surprised

[6] For Hodgkin's take on the relevance of membrane resistance, see Hodgkin (1994, pp. 130–131).

[7] Hodgkin (1994, pp. 130–135).

[8] Hodgkin (1994, p. 133) credits Huxley with the idea; Huxley (2004, p. 292) credits Hodgkin with the idea.

to find that the action potential exceeded the resting potential. Hodgkin's earlier experiment with external electrodes in crab had prepared them for this result.[9]

The theory of singular compositional abductive disconfirmation provides an account of HH's implicit reasoning in 1939. If the action potential in one of the axons studied in 1939 were due to a breakdown in the membrane permeability to ions, then its action potential should approach zero. The axon's potential does not approach zero; it overshoots zero. Therefore, the action potential is not due to a breakdown in the membrane permeability. This hypothetical reasoning should be interpreted in terms of abductive disconfirmation. It is reasoning about determination relations among things in the world; it is not merely a set of true sentences constituting an instance of modus tollens.

Given the theoretical significance of their result, the next order of business for HH was to verify the qualitative correctness of the overshoot measurements and then develop an accurate quantitative measure of the action potential.[10] In terms of the framework described in Chapter 4, HH, along with Curtis and Cole, were interested in the extent to which their data provided reliable and accurate indications of their data targets. In other words, they were concerned that their data provided reliable results. This concern was a recurring feature of the physiological research of this time, a feature worthy of its own philosophical treatment. I do not mean to provide an exhaustive treatment of these issues here. I do not even try to treat all the issues HHC&C faced in these experiments. My more modest goal is, first, to present enough of the history to indicate that there was an issue and, second, to provide a concrete motivation in the case at hand for the distinction between data and results. For this purpose, I will only touch on two of HHC&C's concerns: potential axonal damage due to the insertion of microelectrodes and junction potentials. These were prominent features of the HHC&C discussions but are also relatively easy to explain. I will set aside factors concerning the electrical features of the experimental apparatus.[11]

[9] Huxley (2004, p. 292).

[10] Hodgkin and Huxley (1945, p. 36) states this explicitly. Note that Hitler's invasion of Poland in 1939 and the call to British military service prevented HH from further experimentation on the matter until after the war.

[11] For a more robust discussion of the issues, see Hodgkin and Huxley (1939), Curtis and Cole (1940, 1942), and especially Hodgkin and Huxley (1945). For further issues with different data and different data targets, see Hodgkin (1938, pp. 89–92), Hodgkin and Huxley (1947, p. 343), Hodgkin and Katz (1949a, pp. 38–42), Hodgkin and Katz (1949b), and Hodgkin et al. (1952).

One of HHC&C's more obvious concerns was that inserting an electrode into an axon damages the axon. Hodgkin and Huxley (1945) report, for example, that the process of dissecting out an axon and inserting a cannula to carry the internal electrode took about 2 h, but that should the microelectrode touch the axonal membrane, the fiber would be "spoilt."[12] To avoid this, HH developed a sophisticated system of mirrors that would better enable them to guide the insertion of the electrode down the middle of the axon.[13] Curtis and Cole (1940, 1942) and Marmont (1949), for their part, developed a cell whose axis would carry the electrode down the center of the axon.

HHC&C relied on two principal checks on possible damage to the axons. First, they relied on comparisons of features of axons before and after the insertion of the microelectrode. So, for example, Curtis and Cole measured impedance before and after insertion and found only a small percentage variation.[14] HH compared the magnitude of the action potential with an external electrode before and after internal microelectrode insertion.[15]

The second check on injury was the constancy of axonal behavior. In their initial report on intracellular electrodes, HH observed that axons survived and transmitted impulses for several hours.[16] Curtis and Cole comment that "some axons remained excitable for as long as 4 hours after impalement. Usually the end which was punctured became inactive very soon and, starting from that point, the rest of the axon would slowly become inactive" (Curtis & Cole, 1940, p. 151).[17] Later, they comment that "Results obtained from axons in which the potentials did not remain constant for at least an hour after impalement were discarded" (Curtis & Cole, 1942, p. 138).

As something of an aside, notice that compositional abduction is at work in so modest a context as developing a sound experimental technique. Consider the explanation-seeking why-question, "Why did the end of the axon that was punctured become inactive very soon and then, starting from that point, the rest of the axon slowly become inactive?" Schematically speaking, HHC&C thought that some activity instances in

[12] Hodgkin and Huxley (1945, pp. 177–179). See also Hodgkin (1994, p. 132) and Hodgkin (1951, p. 340). Note that HH did not directly measure damage to the axonal membrane caused by a microelectrode. Instead, they hypothesized that touching the membrane caused damage since that would explain the electrical behavior of the axon.

[13] Hodgkin and Katz (1949a, p. 38) describes a different mirror system for the same purpose.

[14] Curtis and Cole (1940, p. 151).

[15] Hodgkin and Huxley (1945, pp. 180, 183).

[16] Hodgkin and Huxley (1939, p. 711). See also, Hodgkin and Huxley (1947, p. 343).

[17] See Curtis and Cole (1942).

some individuals in the membrane ceased. The inactivation of the membrane was implemented by a cessation of certain activity instances in these unspecified individuals of the membrane. What caused the cessation of the activity instances of the parts of the membrane was contact with the microelectrode. When HHC&C proposed that an electrode's touching the membrane damages the membrane, they did not have any direct method to measure what the touching did to the membrane. They simply postulated damage to some parts because that would explain the subsequently measured results in the axon. When Curtis and Cole observed that axonal inactivity advanced from the point of microelectrode insertion, they inferred that the insertion damaged the axon's parts because that would explain the inactivity moving from the insertion point down the axon.

Another of HHC&C's concerns, arising later, focused on junction potentials. This was a potential source of systematic error that had a greater bearing on the quantitative features of the potentials than on the qualitative overshoot of the action potential.[18] Junction potentials arise from the migration of ions at the interface between solutions of different ionic constituents. So, Curtis and Cole (1942) note two junction potentials: one between the KCl of the microelectrode and the seawater and the other between the axoplasm and the KCl of the microelectrode. The former could be accurately calculated from the known ionic constituents, but the latter could not, because of the uncertain ionic concentration of the axoplasm. For the latter, only an estimate was available.[19]

The upshot here is that HHC&C gave reasons to believe that what appeared on their oscilloscopes – their data – provided more or less reliable and more or less accurate indications of resting and action potentials – their data targets or results. Having arrived at this conclusion, they began to work on the ramifications of their results. Hodgkin and Huxley (1945, pp. 189–194) considered four rival hypotheses that might explain the overshoot consistent with the membrane theory. Given how foreign these rivals are to contemporary thinking and how quickly HH abandoned them, I will not digress to review them.

5.2 The Transition to a New Theory: The Examination of Potassium

Proceeding on the assumption that something was amiss with the membrane theory, HHC&C found it natural to take a more quantitative look

[18] See Hodgkin and Katz (1949a, p. 37). [19] Curtis and Cole (1942, p. 142).

at the role of potassium ions.[20] Curtis and Cole (1942) measured both resting and action potentials while varying the extracellular concentration of potassium ions. They found that the resting potential changed with KCl concentration as predicted by the Nernst equation (although they did not mention the equation by name).[21]

The Hodgkin and Huxley (1946) investigation of potassium – another brief note in *Nature* – nicely illustrates many of the ideas I have been advancing. Given its brevity, I can review the paper almost line by line. Moreover, conveniently for my discussion, it sets aside the technical details of the experimental apparatus and axon preparation to focus on the experimental results and how HH interpreted them.[22]

Hodgkin and Huxley (1946) was HH's first paper based on postwar experimentation.[23] Rather than pursuing a rival to the membrane theory, HH focused on exploring the consequences of extracellular potassium for action potentials in *Carcinus*. According to the hypothesis, sustained axon activity should release potassium into the extracellular medium. Their goal was to provide "an indirect but very sensitive measure" (Hodgkin & Huxley, 1946, p. 376) of this potassium loss.

To begin with, based on unpublished results, HH believed that if the extracellular KCl content is roughly tripled (from 9.8 to 29.8 mmol/l), then membrane conductance (the inverse of membrane resistance) is roughly tripled. By contrast, trebling extracellular NaCl concentration produces no measurable change in membrane conductance. The NaCl control, thus, eliminates one rival hypothesis concerning the basis for a change in membrane conductance. Therefore, they adopt as a working hypothesis that extracellular KCl increases membrane conductance.

Next, HH proposed, based on optical and electrical measurements, that isolated axons immersed in oil are surrounded by a small volume of seawater adhering to the external surface of the axon.[24] They reasoned hypothetically that, if axonal activity releases KCl into this small space, then sustained tetanic axonal activity should release enough KCl into this space to produce a measurable change in membrane conductance. They then tested this consequence of the KCl-release hypothesis by stimulating

[20] Hodgkin (1976, p. 13).

[21] Curtis and Cole (1942, pp. 139, 143). Curtis and Cole measured a 50 mV increase in resting potential with a tenfold increase in extracellular potassium. This was close to the "theoretical" value of 58 mV specified by the Nernst equation. See Hodgkin (1994, p. 273) or Brown (2020, p. 24).

[22] These details are, however, discussed in Hodgkin and Huxley (1947).

[23] Hodgkin and Huxley (1945) and Hodgkin and Rushton (1946) were write-ups of prewar work.

[24] For details of HH's calculation of this volume, see Hodgkin and Huxley (1947, pp. 354–355).

an oil-immersed axon for 1 min, and then measuring the conductance. As predicted, the conductance increased. This is an instance of abductive confirmation of the hypothesis that action potentials release intracellular potassium.

HH also reasoned abductively about another result. They noted that, within a few minutes after the tetanic stimulation, the conductance returned to close to its pre-stimulation value. Given this return, HH proposed that the axon rapidly reabsorbs the released potassium. The reabsorption hypothesis would explain the return of the conductance to the neighborhood of its pre-stimulation value.

HH also considered a rival to the potassium reuptake hypothesis: The effects "might equally well have been due to some structural alteration in the membrane which did not depend upon a chemical change in the external medium" (Hodgkin & Huxley, 1946, p. 377). HH performed a control experiment to rule this out. Following the minute of simulation, they immersed the axon in seawater for a few seconds. Their thinking was that this would clear away the extracellular potassium that had accumulated in the extracellular region delimited by the oil. The result was that the axon conductance very quickly returned to the pre-stimulation values. Here we have compositional abductive disconfirmation of the "structural alteration hypothesis."

Having concluded that axonal activity releases potassium, HH then turn to a computation of the amount of potassium released. This is a simple calculation based on the volume of the aqueous solution clinging to the axon and the change in conductance. Based on eleven experimental determinations, they computed an average of 1.7×10^{-12} moles/cm^2 per impulse. Notice that HH were interested in the individual, spatiotemporally localized determinations, since those individual determinations were used to compute the average. This average value can also be checked against values of the membrane capacity. Notice that this computation only supports their conclusion that the axon releases 1.7×10^{-12} moles/cm^2 per impulse on the assumption that action potentials really do release potassium. This is grounds for thinking that HH took the abductive reasoning to lead to a true conclusion, that the abductive reasoning provides confirmation.

HH's experimental results raised the explanation-seeking why-question, "Why did the membrane conductance of the oil-immersed axon increase after prolonged stimulation?" Their explanation is that axonal activity releases potassium into the limited extracellular space and that this increase in extracellular potassium raises the membrane conductance. The compositional hypothesis that potassium efflux implements axonal activity, in part, explains the experimental results. It is HH's interpretation of their experimental results.

Hodgkin and Huxley (1947), sharing the same title as Hodgkin and Huxley (1946), elaborates that earlier work, including details about the apparatus and preparation. It includes another abductive inference based on downstream consequences of axonal activity.[25] HH propose that there should be a linear relationship between the frequency of the tetanic stimulation of the axon and the conductance change, on the assumptions that each action potential releases the same amount of potassium and that conductance is proportional to external potassium concentration. Upon performing the requisite experiment, HH did find a roughly linear relationship. Again, there is support for the hypothesis that action potentials release intracellular potassium since this hypothesis would be part of the explanation of the linear relationship between the frequency of stimulations and the conductance changes.

5.3 The New Theory

In Hodgkin's recounting of the matter, Huxley had assumed that sodium must play a role in the axonal potentials, whereas he was skeptical.[26] What prompted Hodgkin to rethink the matter was receiving a paper from Katz in late 1946 showing that crab axons were inexcitable in salt-free sugar solutions.[27] In January 1947, Hodgkin showed that, in crabs, reducing extracellular sodium reduced the action potential. This prompted Hodgkin to do more sophisticated experiments on squid axons in the summer. After a long delay, the results of these experiments, in time conducted with Katz, made their way into Hodgkin and Katz (1949a).[28] Here, I will not provide an exhaustive review of the lengthy (Hodgkin & Katz, 1949a), but will instead focus on the "Overton experiment," a replication of an experiment performed decades before in plants by Charles Overton.[29] I limit myself to these experiments, as they illustrate singular compositional abduction in controlled experiments. This is perfectly consistent with my proposal that singular compositional abduction plays an important, even if not exclusive, role in Hodgkin and Katz's reasoning.

[25] For still another related experiment, see Hodgkin and Huxley (1947, pp. 352–353).
[26] Hodgkin (1976, p. 14) and Hodgkin (1994, pp. 269–270). [27] Katz (1947).
[28] Hodgkin (1994, p. 277). [29] Overton (1902).

5.3.1 *The "Overton Experiment"*

Chapter 3 proposed that singular compositional abductive inferences implicate explanations of spatiotemporal particulars. Spatiotemporal particulars are especially salient among the first results reported in Hodgkin and Katz (1949a). For one thing, Hodgkin and Katz report that they began their experiments with a series of measurements, which they report in their table 3 (reproduced in part in Table 5.1). Each entry in this table reports a spatiotemporal particular, such as the resting potential of an individual axon at a particular time and place. Based on data collected with a thermometer, silver–silver chloride microelectrodes, oscilloscope, etc., these results are then available for further inferences. Hodgkin and Katz also provide averages of the results, such as the average resting potential of the axons. The individual results, perhaps accompanied by elementary mathematical computations, can serve as the basis of empirical generalizations, such as, "All shore crab axons have a resting potential of 48 mV." The point I emphasize here is that spatiotemporal particulars have a role to play in Hodgkin and Katz's reasoning: they provide the values upon which

Table 5.1 *Electrical properties of axons in seawater*

Temperature	Resting potential	Action potential	Maximum rate rise	Maximum rate fall
(°C)	(mV)	(mV)	(V.sec^{-1})	(V.sec^{-1})
22	46	85	-	-
20	52	93	-	-
21	52	86	490	290
24	51	83	580	380
22	50	86	650	400
22	49	93	770	460
20	40	87	560	330
20	51	98	630	390
–	48	87	520	340
21	46	89	600	360
20	53	99	1,000	530
20	46	85	620	480
19	42	85	490	330
20	45	86	590	360
21	45	82	680	350
Average				
21	48	88	630	380

the averages are computed and they provide the values from which a scientist might draw empirical generalizations.[30]

These points survive the observation that each entry in the table may only be a selection of one of perhaps a handful of measurements Hodgkin and Katz took of each axon. Even if Hodgkin and Katz selected only one of many measurements, they did select one spatiotemporal particular to report. This point also survives the possibility that HH were reporting an average of multiple measurements of a single axon resting potential, for even in that case, each individual measurement would be of a spatiotemporal particular that was used to compute the average for that axon. Finally, these points survive the contention that empirical generalizations are typically more important to scientists than mere mathematical averages.

Hodgkin and Katz's table reveals that they were concerned with spatiotemporal particulars, but there is another spatiotemporal particular that was of great interest to them. They write,

> [W]e have never observed action potentials greater than 100 mV. at 18–23°C., whereas Curtis & Cole describe a spike as large as 168 mV. in a fibre which gave a resting potential of 58 mV. The matter is not one that can be lightly dismissed, because the existence of a fibre capable of giving an overshoot of 110 mV. has far-reaching implications. (Hodgkin & Katz, 1949, p. 43)

Here, Hodgkin and Katz were clearly concerned with a single action potential recorded by Curtis and Cole. It had a spike of 168 mV. Why were they interested? Because they believed that this result was in some sense anomalous. If correct, it would have "far-reaching implications." In the end, Hodgkin and Katz did not explain the anomaly, but that is not the point I need to emphasize. Instead, it is that the consideration of spatiotemporal particulars has some role to play in scientific reasoning. Often, this role is understated as scientists seek to generalize, but they nevertheless have a role to play. The theory of singular compositional abduction is intended to explicate what this role is sometimes like.

Section 3.4 introduced the Overton experiment. Recall that the gist of the experiment was to measure the action potential of a squid axon over time. It began with the axon in a sodium-containing seawater solution,

[30] Table 3 in Hodgkin and Katz (1949a) also represents spatiotemporal particulars, but does not provide averages. So, that table does not as nicely highlight the differences between particulars, on the one hand, and computations and generalizations from particulars, on the other. Table 2 of Hodgkin and Huxley (1952a, p. 464) provides both individual measurements and averages.

then switching to a sodium-free medium (isotonic dextrose), then switching once again to a sodium-containing seawater solution. In Section 3.4, I reviewed Hodgkin and Katz's interpretation of two principal results of this experiment. First, after the seawater was removed, the peak of the action potential declined, but once the seawater was restored, the peak of the action potential returned to near-normal levels. Second, after the seawater was removed, the rate of rise of the action potential decreased, but once the seawater was restored, the rate of rise of the action potential returned to near-normal levels.

After the core Overton experiment, Hodgkin and Katz explored various controls. These include experiments with reduced-sodium media and sodium-enriched media. For simplicity, the present discussion will focus on the reduced-sodium experiment. In this experiment, Hodgkin and Katz used a sodium, reduced-sodium, sodium protocol with temporal delays (typically about 15 min) between measurements. Hodgkin and Katz assumed that these delays allowed the axons to equilibrate to the changed ionic concentrations. The three reduced-sodium media contained (1) 33% seawater, 67% isotonic dextrose; (2) 50% seawater, 50% isotonic dextrose; and (3) 71% seawater, 39% isotonic dextrose. The experiment showed that the peak of the action potential was reduced with greater reductions in the sodium concentration of the medium. Further, they showed that the effects of the reduced sodium were reversible. Still further, the resting potentials were largely unchanged. The philosophical analysis just given of the first experiment easily carries over to this experiment.

5.3.2 "Direct" Measures of Ion Movements

Although my focus has been, and later will be, on the electrophysiological methods used by Cole, Curtis, Hodgkin, Huxley, and Katz, it is useful to draw attention to other methods that provided background for the electrophysiological work and were important at the time. These are methods that are sometimes omitted in contemporary textbook treatments.

Starting in 1946, Keynes began a series of experiments with a recently introduced technique of radioactive labeling, which enabled him to use radioactive K^+ and Na^+ to trace the movements of these ions in and out of cells. For present purposes, I will focus only on the work on potassium, but essentially the same philosophical account will apply to the experiments on sodium.

Keynes (1948) – a brief communication in the *Proceedings of the Physiological* Society – reported soaking *Carcinus* nerves in a Ringer's solution containing radioactive K^{42} for several hours.[31] Keynes assumed that during this period radioactive potassium would collect within the cell. He then transferred the nerve to a bath that circulated non-radioactive Ringer's solution, thereby carrying away any K^{42} that left the cell. Keynes found that over time the amount of radioactivity in the axon, as measured with a Geiger counter, decreased exponentially. In addition, when the nerve was stimulated at 17 impulses/sec, the level of intracellular radioactivity decreased more rapidly.

Consider what is going on in this simple experiment in light of the theory of Chapter 3. Consider, first, data and results. The Geiger counter generates data that is aimed at counting the number of nuclear decay events per unit time. Each datum on the counter is a spatiotemporal record aimed at another spatiotemporal event, a nuclear decay. Each decay is an activity instance of an individual. It is something that an individual K^{42} atom does just once. Keynes and the other scientists of his day presumed that past research had validated the assumption that the data on the Geiger counter delivers genuine results about decay events.

Next, consider the explanation of rates. Keynes observed that the amount of radioactivity in the axon decreased exponentially over time. He explained this in terms of the number of radioactively labeled potassium ions leaving the axon and being washed away by the circulating Ringer's solution.[32] The exponential decay over time was due to the potassium ions leaving the cell in proportion to their concentration. He further explained the greater decrease in radioactivity during axonal activity in terms of the release of more potassium ions during axonal activity.

Third, consider the explanation of the results of controlled experiments. In Keynes's experiments, the rate of decay under stimulation of the nerve is greater than the rate of decay under resting conditions. One rate is greater than another. This is a comparison that takes place in the context of a controlled experiment. This invites the question, "Why is the rate of decay under stimulation greater than the rate of decay under resting conditions?" Answer: More potassium is released when the nerve is stimulated than when it is in a resting state. This is an explanation in terms of the relative

[31] See also Keynes (1949), and Keynes and Lewis (1951a, 1951b). Rothenberg (1950) included studies of radioactive sodium.

[32] Another factor in the decrease would have been a small decrease in the radioactivity of the K^{42}. It would have been small, since the half-life of K^{42} is about 12.5 h and the experiment ran for over 3 h. Keynes (1948) does not discuss this.

numbers of activity instances of individuals. Keynes did not spell all of this out in his paper, since it was evident to the physiologists at the time.

Mid-century physiologists perceived a difference between the HHC&C electrophysiological methods and Keynes' radioactive tracer methods. They conceptualized this difference in terms of "direct" and "indirect" methods.[33] Hodgkin (1951) offers these comments on the differences between electrophysiological methods and chemical methods:

> [T]he introduction of radioactive tracers and the development of sensitive chemical methods have enabled ionic movements to be followed experimentally. These methods are direct, in the sense that they leave no doubt about the identity of the ion under observation, but they suffer from the disadvantage that their time-resolving power is very poor. They can, for example, be used to prove that potassium ions leak out of a nerve when it is active, but they do not show whether this is a sudden event which forms an essential link in the process of conduction or whether it is a relatively slow affair connected with the recovery process. The resolving power of electrical instruments is now much better than anything that physiologists are likely to require, but the results are indirect, in that they give no certain information about the identity of the ions which carry the electric current. (Hodgkin, 1951, p. 339)[34]

Hodgkin notes that the radioactive tracer experiments show that sodium and potassium cross the membrane, but they do not show the precise time course. The way to read Hodgkin's comment about "no certain information" is as a recognition that the identity of the ions carrying the current is hypothetical. It is given by compositional abduction.

5.4 Hodgkin and Huxley 1952

As mentioned in the introduction of this chapter, my treatment of the 1952 Hodgkin–Huxley papers will be selective.[35] To describe matters informally, the Hodgkin–Huxley papers offer three "stories." First, there is a driving force story, which is an account of the potential and

[33] See, for example, Keynes and Lewis (1951a).

[34] Just to be clear, Hodgkin (1951), being a lengthy review article, contains a lot of information that informed Hodgkin, Huxley, and Katz's thinking. But, as a review article, rather than an experimental article, it does not fall within the scope of my case for singular compositional abduction.

[35] Brown (2020) and Raman and Ferster (2022) provide extensive exposition and commentary on the entire series of papers. Aidley (1998, pp. 58–65) provides a helpful complement to the present account. It is a more compact textbook presentation, albeit sometimes anachronistic, that captures the course of HH's experimentation as described here. Hodgkin (1994, pp. 288–296) also provides a concise exposition of the course of the experimentation described here.

concentration gradients that generate forces that can move ions across an axonal membrane. Second, there is a permeability story. A simple lipid bilayer membrane is largely impermeable to ions. Yet, at rest, the membrane is apparently selectively permeable to some ions and, during an action potential, this permeability rapidly and selectively changes. Finally, there is a propagation story. This is a story about how local changes at one point along an axon induce changes in adjacent regions of the axon so as to propagate the action potential into those adjacent regions.

From the experimental work in 1947, along with the radioactive tracer work, HH had by 1949 embraced the sodium–potassium understanding of the action potential. This substantiated part of the driving force story. In the fall of 1947, HH heard from Cole about two experimental techniques. The first provided for feedback control which enabled an experimenter to hold the current constant or the potential across the membrane at a desired command potential. The voltage clamp enabled the study of another part of the driving force story. The second involved using long electrodes that could maintain a constant voltage along a substantial length of the axon. The space clamp eliminated complications that arise with the generative feedback underlying the propagation of the action potential, thereby enabling HH to develop the driving force and permeability stories apart from the propagation story. In March of 1948, Hodgkin spent several days meeting with Cole and George Marmont, learning in person about these techniques.[36] One of the central conclusions Hodgkin took away was that by using feedback to maintain a constant voltage, one observes that a 50 mV depolarization is followed by a brief surge of what he interpreted as a capacity current, then a short phase of inward current, and then finally an outward current.

After spending more than a year refining their experimental technique, Hodgkin, Huxley, and Katz spent much of the summer of 1949 performing the experiments that, after various delays, formed the basis for the five classic papers of 1952. Of these papers, the second, third, and fourth papers, that is, Hodgkin and Huxley (1952a, 1952b, 1952c) will be most relevant. The first paper – Hodgkin et al. (1952), the "voltage-clamp paper" – focuses on the description and validation of the experimental procedures, whereas the fifth paper – Hodgkin and Huxley (1952d), the "modeling paper" – works the experimental results into a quantitative description.

[36] Hodgkin (1976, p. 17) and Hodgkin (1994, p. 282).

5.4.1 "Currents Carried by Sodium and Potassium"

HH distinguish three theoretical goals for Hodgkin and Huxley (1952a), the "Currents Carried" paper. They are: (1) the study of the role of sodium concentration in the external medium, (2) the resolution of ionic currents (a hypothetical contributor to the total current) into sodium and potassium currents, and (3) the expression of membrane permeability in terms of units of ionic conductance.[37] These three goals form the basis for the following three subsections.

The effects of sodium concentration. As I have repeatedly referred to the first experiment of Hodgkin and Huxley (1952a) as illustrating one or another feature of abductive inference, there is no need to review that again here. Section 3.3 above introduced HH's second experiment highlighting the role of the potential across the membrane. Recall from Section 3.3 that, in this second experiment, HH retained the sodium-containing, sodium-free, sodium-containing protocol, thereby retaining experimental control over the sodium in the external medium. In addition, the experiment varies the potential at which the axon is clamped, thereby expanding the number of variables under experimental control. Assuming the reliability of the oscilloscope, an oscilloscope trace (a datum) in one run of this experiment provides evidence for an initial inward current in axon no. 21 (a result). Further, it could be shown that these currents are slowly reduced and eventually replaced by an initial outward current, as the voltage clamp potential is decreased. Each trace corresponds to a numerically and qualitatively distinct current in axon no. 21 and each current may be given a singular compositional explanation in terms of ion fluxes. This family of singular compositional explanations supports the hypothesis that the current flowing across the membrane is the resultant of two opposing forces, a chemical gradient and a potential gradient.

The following comment from HH is one piece of textual evidence supporting a role for abduction in the interpretation of the first two experiments:

> When the axon is placed in a sodium-free medium, such as the 'choline sea water', there can be no inward flux of sodium, and the sodium current must always be outward. This will account for the early hump on the outward current which is seen at all but the lowest strength of depolarization in the centre column of Fig. 2. (Hodgkin & Huxley, 1952a, p. 452)

[37] Hodgkin et al. (1952, p. 425).

The first sentence draws attention to the sodium concentration gradient. It was noted earlier that there is sodium inside the axon, so that in a sodium-free medium, the concentration gradient would run from higher in the axon to lower in the extracellular medium. Physiologists also knew that sodium would flow down this concentration gradient, unless this gradient was offset by a counteracting potential gradient. So, supposing that "account for" is synonymous with "explain," HH are proposing that the "early hump" in the outward current is an explanandum that is compositionally explained by the sodium concentration gradient.

Later sections of Hodgkin and Huxley (1952a) indicate that HH understand their abductive inferences to confirm the sodium hypothesis. At an early point in their discussion section, HH comment, "We have shown in the earlier parts of this paper that there is good reason for believing that the component of membrane current that we refer to as I_{Na} is carried by sodium ions" (Hodgkin & Huxley, 1952a, pp. 465–466). This is connected to confirmation by assuming that if something provides good reason for P, then it provides confirmation for P. A second comment, in the final summary section, adds another piece of textual evidence in support of the view that HH think abduction provides confirmation. HH review the first two experiments and then comment, "These results support the view that depolarization leads to a rapid increase in permeability which allows sodium ions to move in either direction through the membrane" (Hodgkin & Huxley, 1952a, p. 471). Supposing that "support" is synonymous with "confirm," the idea is that the sodium gradient's explanation of the early hump confirms the sodium hypothesis. Contrast what HH write with what they do not write. They do not write that, "In the earlier parts of this paper we have introduced the hypothesis that the component of membrane current that we refer to as I_{Na} is carried by sodium ions." They do not write that, "In the earlier parts of this paper we have proposed that the hypothesis that the component of membrane current that we refer to as I_{Na} is carried by sodium ions is pursuit-worthy." So, in adopting the sodium hypothesis because it is compositionally explanatory, HH do not intend merely to introduce the sodium hypothesis as worthy of further investigation. They intend to confirm it, to provide good reasons to believe it, or to support it. So, while the abductive interpretation of HH's work is, obviously, an interpretation, there are textual grounds for it. It bears emphasizing that HH's comments on their work complement the direct examination of their work.

As part of their interpretation of their second experiment, HH note that, "These results are in *qualitative* agreement with the hypothesis that the inward current is carried by sodium ions" (Hodgkin & Huxley, 1952a, p. 450, italics added). Further, HH note that, in the case of the medium containing sodium, there is a "critical value" – the sodium potential, E_{Na} – at which the inward driving force of the sodium concentration gradient equals the outward driving force of the membrane potential. Further, the sodium potential should be determined according to the Nernst equation. These two comments about the qualitative agreement and the Nernst equation foreshadow things to come in their third experiment, an experiment that provides what they take to be quantitative confirmation of the Nernst equation specification of the sodium potential.

HH's third experiment is a variation on their second, wherein the sodium-free solution is replaced by a sodium-diluted solution. This is a successor to the "reduced-sodium experiments" of Hodgkin and Katz (1949a). One sodium-diluted solution contains 10% of the sodium of seawater; the other 30%. Further, the sodium-containing, sodium-free, sodium-containing protocol is slightly modified in favor of a reduced-sodium-containing, sodium-containing, reduced-sodium-containing protocol.

Before beginning to interpret their results, HH note that the figures they provide do not represent the total current. Instead, they represent a hypothetical ionic current. HH postulated that the total current is the sum of a capacity current and an ionic current, that is, $I_{total} = I_c + I_i$. HH hypothesize that the potential change associated with voltage clamping gives rise to a very brief capacity current. This current can be measured for a single case, then scaled appropriately for different potential changes, and then subtracted from the measured total currents to yield a hypothetical ionic current. Recall a point from Chapter 2 regarding derivative compositional explanations. Here is a case in which HH were able to provide more than a mere derivative compositional explanation of the total current, they were able to provide a compositional explanation of the capacity current, which is a contributor to the total current.

Notice, in addition, that HH's comment about the capacity current reveals implicit compositional abductive reasoning. They measured a brief spike in total current lasting about a microsecond. It is the time course of this spike just after the onset of the voltage clamp that led HH to this hypothesis. They had no experimental means of directly detecting the congregation of charges on either side of the axonal membrane, as was hypothesized in a capacity current. By contrast, a capacity current would explain the brief time course of this spike following the onset of the voltage clamp, thereby giving HH reason to believe that a component of the total current is due to a capacity current. HH's fellow physiologists would surely comprehend this reasoning, albeit not necessarily described in terms of abductive reasoning.

HH report that the results of their third experiment are in keeping with what one would expect from the second experiment.

> Inward current is present, but only over a range of membrane potentials which decreases with the sodium concentration, and within that range, the strength of the current is reduced. A definite sodium potential still exists beyond which the early hump of ionic current is outward, but the strength of depolarization required to reach it decreases with the sodium concentration. (Hodgkin & Huxley, 1952a, p. 452)

The more significant aspect of this experiment, however, is presented in a separate section, "The external sodium concentration and the 'sodium potential'." This section begins with the claim, "Estimation of the 'sodium potential' in solutions with different sodium concentrations is of particular importance because it leads to a quantitative test of our hypothesis" (Hodgkin & Huxley, 1952a, pp. 452–454). This "quantitative test" is an instance of extended hypothetical reasoning.

HH note that, just as the Nernst equation determines the sodium potential in seawater, so it determines corresponding sodium potentials, E'_{Na}, in the two reduced sodium media:

$$E'_{Na} = \frac{RT}{F} \log_e \frac{[Na]_i}{[Na]'_o}.$$

They next note that the difference between these two sodium potentials – a sodium potential shift – is given by

$$E'_{Na} - E_{Na} = \frac{RT}{F} \left\{ \log_e \frac{[Na]_i}{[Na]'_o} - \log_e \frac{[Na]_i}{[Na]_o} \right\} = \frac{RT}{F} \log_e \frac{[Na]_o}{[Na]'_o}. \quad (1)$$

E_{Na} and E'_{Na} yield displacement membrane potentials given by $V_{Na} = E_{Na} - E_r$ and $V'_{Na} = E'_{Na} - E'_r$, where E_r is the resting potential in seawater and E'_r is the resting potential in a reduced-sodium medium. Rearrangement and substitution of these last equations yields

$$(V'_{Na} - V_{Na}) + (E'_r - E_r) = \frac{RT}{F} \log_e \frac{[Na]_o}{[Na]'_o}. \quad (2)$$

(2) yields an "observed shift" specified by

$$E'_{Na} - E_{Na} = (V'_{Na} - V_{Na}) + (E'_r - E_r) \quad (3A),$$

Table 5.2 *Observed and theoretical change in sodium potential when the extracellular medium is changed from seawater to a low sodium medium*

Axon no.	Temp. (°C)	$\frac{[Na]'_o}{[Na]_o}$	V_{Na} (mV)	V'_{Na} (mV)	(E'r - Er) (mV)	Observed (mV)	Theoretical (mV)
20	6.3	0.3	-105	-78	+3	+30	+28.9
20	6.3	0.1	-96	-45	+4	+55	+55.3
21	8.5	0.1	-100	-48	+4	+56	+55.6
21	8.5	0.1	-95	-45	+4	+54	+55.6

and a "theoretical shift" specified by

$$E'_{Na} - E_{Na} = \frac{RT}{F} \log_e \frac{[Na]_o}{[Na]'_o} \qquad (3B).$$

All terms in these last two equations can be measured, thereby enabling HH to compare the theoretical shift with the observed shift.[38] This comparison appears in Table 5.2, reproduced from HH's table 1. Commenting on this comparison, HH write,

> It will be seen that there is good agreement, providing strong evidence that the early rise or fall in the recorded ionic current is carried by sodium ions, moving under the influence of their concentration difference and of the electric potential difference across the membrane. (Hodgkin & Huxley, 1952a, p. 454)

HH's two brief paragraphs invite two philosophical observations. First, historians and philosophers of science should step back from this relatively complicated computation to consider a point that HH do not make, but that would have been evident to mid-twentieth-century physiologists. In point of logic, one might have tested the application of the Nernst equation by substituting values into the right-hand side and comparing the results with the measured results. Although this test was logically possible, it was practically impossible for HH. To accurately measure $[Na]_i$ using the methods available at that time would have damaged the cells under study. Here is Jane Cronin's commentary on HH's "indirect procedure" based on the third experiment:

[38] Strictly speaking, the V_{Na} was not measured, but interpolated as spelled out in Hodgkin and Huxley (1952a, pp. 454–455).

> [To test the sodium potential hypothesis], the reasonable procedure would seem to be to compute E_{Na} by using the Nernst formula and to compare the observed experimental value (close to 79 mV) just described. The difficulty in such a direct procedure is that the value of $[Na]_i$ is needed, but $[Na]_i$ is not known with any accuracy for the axon being studied experimentally. (Indeed, to measure $[Na]_i$ would require the destruction of the axon.) Consequentially, . . . the more indirect procedure is used. (Cronin, 1987, p. 28)[39]

Cronin's commentary draws attention to what would have been well-known to physiologists at the time. Moreover, it explains why HH used the relatively complicated experimental method they used in their third experiment. The methodological moral for historians and philosophers of science: limiting one's philosophical account to what scientists make explicit may omit important details. One needs to be aware of relevant background information.

The second philosophical point to note is HH's use of hypothetical reasoning. The hypothesis was that the Nernst equation specifies the relationship between the internal and external sodium concentration and the sodium potential. Because of the technological limitations on measuring the value of $[Na]_i$, HH had to resort to a more complicated experimental procedure, one that required more reasoning than many other examples. In practice, HH had to engage in a more complicated form of hypothetical reasoning. In this reasoning, (3A) gives a hypothesis that is tested against "observations" generated by (3B).

For decades, historians and philosophers of science looking at HH's hypothetical reasoning would have been tempted to interpret it as an instance of hypothetico-deductive confirmation. HH logically deduced an "observable" consequence of their hypothesis, and then noted that this consequence obtained. An abductive theory of confirmation, however, offers a different interpretation of HH's reasoning. What HH's reasoning shows is that the expression in (3B) represents the way in which the values of two concentrations of external sodium ions, etc., will determine two distinct sodium potentials, that is, a sodium potential shift. (3B) represents an ontological dependence relation among things in the world. (3A) reveals another ontological dependence, one that, in combination with measurements of V_{Na}, V'_{Na}, and $(E'_{Na} - E'_r)$, allows a measurement of the sodium potential shift. What is crucial for the abductive approach is that what is

[39] Aidley (1998, p. 59) and Raman and Ferster (2022, p. 79) make the same point.

represented in (3B) explains the sodium potential shift, so that HH believe they have "strong evidence" of what is represented in (3B).

All of this is a more complicated version of the idea broached in Chapter 4. Recall that Chapter 4 included a review of a compact one-paragraph argument from Curtis and Cole (1940). This was the argument that a measured potential change was due entirely to the passage of an action potential. It was an argument that, on its face, serves as a perfect illustration of HD confirmation in action. The reasoning was laid out as an instance of affirming the consequent. Chapter 4 proposed that the theory of HD confirmation provides one philosophical interpretation of Curtis and Cole's reasoning, but that the theory of abductive confirmation provides another. Those who believe that scientists are interested in what is going on in the world, rather than in logical relations among sentences, might incline to the abductive approach. Further support for the abductive approach is that it provides a diagnosis of the tacking problems that have beset HD confirmation for decades. There is, therefore, reason to adopt an abductive interpretation. The hypothetical reasoning of Hodgkin and Huxley (1952a, pp. 452–454) is like Curtis and Cole's, only more complicated.

One final wrinkle. Note that the explanandum in this experiment is the shift in the sodium potential. This is a difference between two sodium potentials, one in seawater and another in a reduced-sodium medium. Each of the two potentials is compositionally explained by way of the Nernst equation and each of these explanations is involved in the explanation of the difference. So, strictly speaking, the proposal here is not that HH's hypothetical reasoning is an instance of compositional abduction. It is more complicated than that, the complication lying in the explanation of a shift. *Prima facie*, this is just an instance of explaining the results of a controlled experiment wherein the results are potentials.

The resolution of ionic currents. Recall that when HH voltage-clamped an axon, they could only measure the total current. They hypothesized, however, that the total current at any given time was the sum of a capacity current and an ionic current. In Hodgkin and Huxley (1952a), they used the time course of the capacity current and its temporal relation to the change in membrane potential to support the capacity current hypothesis.[40] This allowed them to subtract the capacity current from the total current leaving

[40] In Hodgkin and Huxley (1952b), they used a slightly different method involving depolarization and hyperpolarization. This method invoked the symmetry of the resulting currents as a basis for compositionally abductively inferring that there exists a capacity current.

the ionic current, a current that is maintained for much longer than the capacity current. Their next theoretical step was to decompose the remaining ionic current into the currents of different ionic species. For simplicity, I focus on the separation of the hypothetical ionic current into sodium and potassium currents.[41] The arc of their argument involves hypothetical reasoning.

HH propose to separate the sodium and potassium currents using the results of the experiments on currents in seawater and in reduced-sodium media, given certain assumptions.[42] These assumptions – these hypotheses – are the following:

> (1) The time course of the potassium current is the same in both cases.
>
> (2) The time course of the sodium current is similar in the two cases, the amplitude and sometimes the direction being changed, but not the time scale or the form of the time course.
>
> (3) $dI_K/dt = 0$ initially for a period about one-third of that taken by I_{Na} to reach its maximum.
>
> (Hodgkin & Huxley, 1952a, p. 457)

In more detail, (1) is the assumption that differences in the external sodium concentration do not change the potassium current; (2) involves several points we might separate. It says that the external sodium concentration affects the driving force on sodium, which changes the amplitude of the sodium current. Further, with sufficiently large changes in the external sodium concentration, the driving force can be changed from driving sodium into the axon to sometimes driving it out. These changes in the driving force, however, do not change how long the currents flow, that is, the time scale of the current. Nor do these changes in the driving force change the initial rise in inward current or the subsequent decline in the inward current, that is, the time course of the current. Assumption (3) reflects the putative delay in the rise of the potassium current. More specifically, the potassium current is unchanged for the first third of the time it takes for the sodium current to reach its peak.

The procedure for separating the currents began with currents measured using the sodium, low-sodium, sodium protocol and then plotted on a graph. Capacity currents, corrected for the different sodium

[41] To follow up on HH's treatment of other ionic species, one may begin with the single paragraph on Hodgkin and Huxley (1952a, p. 461).

[42] Cronin (1987, pp. 29–34) provides a complementary, more mathematical exposition, of HH's reasoning.

concentrations, were then subtracted from the total current curves. The currents from the first and third curves were then averaged to allow for the deterioration of the axons over the course of the experiment. A further correction was added for the difference in resting potential in the sodium and low-sodium media. They next let I_i, I_{Na}, and I_K represent the ionic current, sodium current, and potassium current in the sodium medium and I'_i, I'_{Na}, and I'_K represent the ionic current, sodium current, and potassium current in the reduced-sodium medium. This permits some simple calculations.

Although not listed among the three assumptions, HH's working hypothesis is that $I_i = I_{Na} + I_K$ and that $I'_i = I'_{Na} + I'_K$. So,

$$I_i - I'_i = (I_{Na} + I_K) - (I'_{Na} + I'_K) = (I_{Na} - I'_{Na}) + (I_K - I'_K). \quad (4)$$

Given assumption (1) that $I_K = I'_K$, it follows from (3) that

$$I_i - I'_i = I_{Na} - I'_{Na} \quad (5)$$

Given assumption (2), that the time course of I_{Na} and I'_{Na} are the same, we have it that $k = I'_{Na}/I_{Na}$, for a constant k. So, $I'_{Na} = kI_{Na}$, so

$$I_i - I'_i = I_{Na} - I'_{Na} = I_{Na} - kI_{Na} = (1 - k)\, I_{Na}. \quad (6)$$

Hence

$$I_{Na} = (I_i - I'_i)/(1 - k) \quad (7)$$

$$I'_{Na} = k(I_i - I'_i)/(1 - k) \quad (8)$$

$$I_K = I'_K = I_i - I_{Na} = (I'_i - kI_i)/(1 - k). \quad (9)$$

(7), (8), and (9) allow plots of I_{Na}, I'_{Na}, I_K, and I'_k in terms of the known quantities I_i and I'_i. These plots are shown in Figure 5.1, redrawn from Hodgkin and Huxley (1952a, p. 459, figure 5).

Were Hempel to review this case, he might interpret HH's reasoning here as another illustration of HD confirmation. He might suggest that HH introduced hypotheses, then derived some consequences from them and then checked those consequences against "observations." Yet, the very nature of the experimental situation precludes comparing the derived sodium and potassium curves from Figure 5.1 with "observed" sodium and potassium curves. The very nature of the experimental situation is that HH could only measure the total current under a voltage clamp, not the isolated sodium and potassium currents. They could not measure the sodium and potassium curves that the HD method prescribes. Consider, now, how this influenced HH's reasoning.

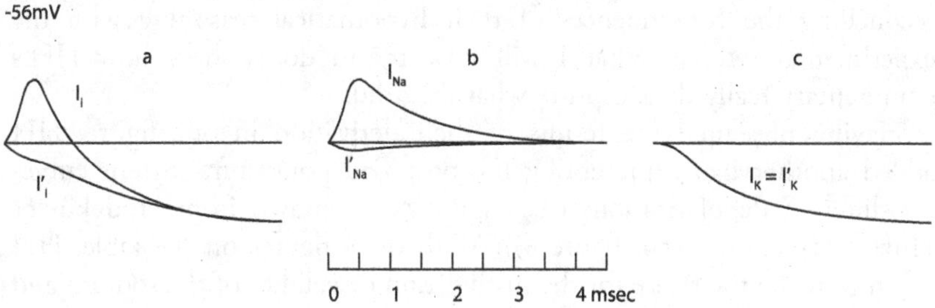

Figure 5.1 "Curves illustrating separation of ionic current into I_{Na} and I_K. a, ionic currents: I_i axon in sea water, membrane potential lowered by 56 mV; I'_i axon in 10% sodium sea water. b, sodium currents: I_{Na} sodium current in sea water; I'_{Na} sodium current in 10% sodium sea water. c, potassium current, same in both solutions." Redrawn from Hodgkin and Huxley (1952a, p. 459, figure 5).

Having introduced their three assumptions, HH immediately provide the following methodological commentary on them:

> The first two of these assumptions are the simplest that can be made, and do not conflict with any of the results we have described, while the third is strongly suggested by the form of records near the sodium potential … These points are sufficient reason for trying this set of assumptions first, but their justification can only come from the consistency of the results to which they lead. (Hodgkin & Huxley, 1952a, p. 457).

Notice that HH introduce the first two assumptions as "pursuit-worthy" because they are simple and because they do not conflict with any of the results they have to date. They do not introduce them because they are explanatory.[43] Also, notice that HH offer some set of hypotheses and refer to "results to which they lead." Given the presuppositions of HD confirmation, one would expect HH to claim that the justification for the hypotheses can only come from their consistency with measurements – with observations. But, of course, given the nature of the experimental situation described in the last paragraph, they cannot check that consistency. They did not have independent access to the sodium and potassium currents; they only had indirect access to the ionic current by abductive reasoning. Moreover, they fully understood this. This is why HH made the slightly different claim that the justification for the hypotheses "can only come from the consistency of the results to which they lead." This interpretation is supported by HH's subsequent reasoning about

[43] These observations do not square with the Peircean line of thought about the introduction of hypotheses.

reconciling the consequences of their hypothetical reasoning with the experimental results. What I will now try to do is show how HH's commentary really does capture what they did.

Having presented the results of their derivation in one figure, HH added another figure presenting the proposed potassium current curves for different depolarizations (see Figure 5.2, redrawn from Hodgkin & Huxley (1952a, p. 459, figure 6)). With these figures on the table, HH begin to make their case for the justification or validity of the sodium and potassium current curves in Figure 5.1:

> A satisfactory feature of these curves, which is to some extent a check on the validity of the assumptions, is that the general shape is the same at all strengths. If the time courses of I_{Na} and I'_{Na} had not been of similar form, Equation [9] would not have removed sodium current correctly. It would then have been unlikely that the curve of potassium current at a potential away from the sodium potential would have been similar to that at the sodium potential, where the sodium current is zero and Equation [9] reduces to $I_K = I_i$ because $k = \infty$. (Hodgkin & Huxley, 1952a, p. 458)

In this passage, HH refer to the potassium curves of Figure 5.2, which all have the same general shape at all strengths. This would not happen if the second assumption, that is, $I'_{Na} = kI_{Na}$, were false. Why? Compare the potassium current with a depolarization to the sodium potential with the potassium current with a depolarization to a potential other than the sodium potential. On the hypothesis that $I_i = I_{Na} + I_K$, (capacity current having been subtracted out), at the sodium potential, the sodium current I_{Na} should be zero, because the driving force is zero, so the $I_i = I_K$. Further, given the proportionality of $I'_{Na} = kI_{Na}$, the other curves should have a shape similar to this. This indicates that HH really did link the justification of their assumptions to the consistency of the results to which they lead.

An abductive interpretation of HH's hypothetical reasoning proposed that the reasoning is a matter of drawing out the ontological consequences of assumptions (1)–(3). The idea is that were things in the world laid out as specified in these assumptions, then the ontological consequences are such that the potassium current curves near the sodium potential should be similar to the potassium current curves far from the sodium potential. Further, those similarities are found to hold. This provides abductive confirmation of the hypothesis, but not textbook HD confirmation.

To conclude this subsection, I remind the reader of a proposal from Chapter 2, that part of Hodgkin and Huxley's achievement was to provide compositional explanations of the capacity current, the sodium current, the potassium current, and the leak current. Further, they showed how

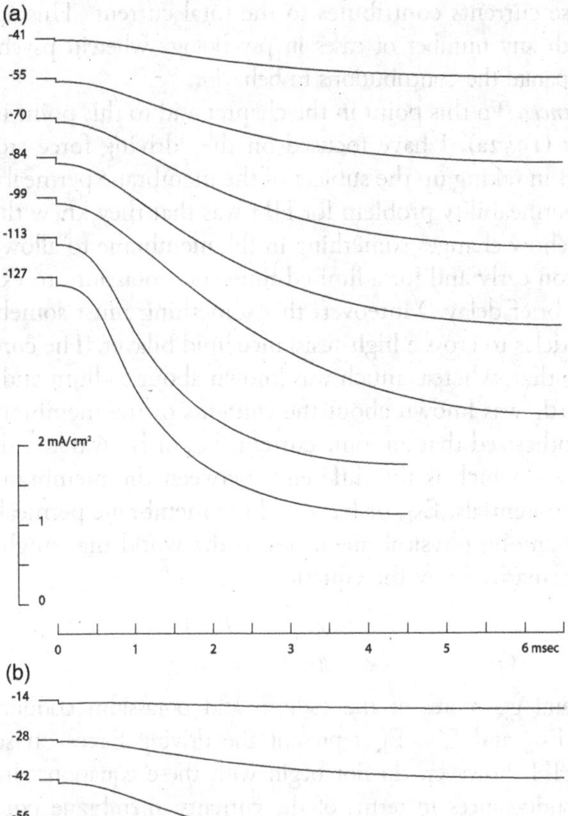

Figure 5.2 Curves of potassium current against time for various strengths of depolarization. Displacement of membrane potential when axon is in seawater is indicated for each curve, in millivolts. a, derived from voltage clamps with axon in 30% sodium seawater, seawater and 30% sodium seawater. Axon no. 20; temperature 6.3°C. b, derived from voltage clamps with axon in 10% sodium seawater, seawater and 10% sodium seawater. Axon no. 21; temperature 8.5°C. Redrawn from Hodgkin and Huxley (1952a, p. 459, figure 6).

each of these currents contributes to the total current. This achievement contrasts with any number of cases in psychology wherein psychologists are unable to separate the contributions to behavior.

Conductances. To this point in the chapter and to this point in Hodgkin and Huxley (1952a), I have focused on the "driving force story." Now, I follow HH in taking up the subject of the membrane permeabilities. The gist of the permeability problem for HH was that they knew that depolarization somehow changes something in the membrane to allow sodium to enter the axon early and for a limited time and potassium to exit the axon following a brief delay. Moreover, this something must somehow enable charged particles to cross a high-resistance lipid bilayer. The core challenge for HH was that, whereas much was known about sodium and potassium ions, very little was known about the contents of the membrane.

HH hypothesized that an ionic current, I_{Na} or I_K, is determined by the driving force – which is the difference between the membrane potential and the ion potentials, E_{Na} or E_K – and the membrane permeability. This is a relation among physical quantities in the world that might be represented mathematically by the equations

$$I_{Na} = g_{Na}(E - E_{Na})$$
$$I_K = g_K(E - E_K),$$

where g_{Na} and g_K represent the sodium and potassium conductances and where $E - E_{Na}$ and $E - E_K$ represent the driving forces on sodium and potassium. HH, however, do not begin with these equations. Instead, they "defined" conductances in terms of the currents, membrane potentials, and ionic potentials:[44]

$$g_{Na} = I_{Na}/(E - E_{Na})$$
$$g_K = I_K/(E - E_K).$$

Based on the results of the third experiment, HH had determined E_{Na} in the way described above. Further, borrowing from their next paper, Hodgkin and Huxley (1952b), they had values of E_K. Finally, given the resolution of the sodium and potassium currents, I_{Na} and I_K, HH could get an empirical handle on g_{Na} and g_K. They could plot the time course for both sodium and potassium conductances (see Hodgkin & Huxley, 1952b, p. 462, figure 8). Further, they were able to determine the rates of rise of conductances against the strength of depolarization.

These equations require both philosophical and scientific commentary. Having introduced the conductance equations, HH comment "the usefulness

[44] See also, Hodgkin and Huxley (1952b, p. 477).

of the definitions, and the degree to which they measure real properties of the membrane, will clearly be much increased if each of these relations is a direct proportionality, so that g_{Na} and g_K are independent of the strength of the driving force under which they are measured. It will be shown in the next paper (Hodgkin & Huxley, 1952[b]) that this is the case" (Hodgkin & Huxley, 1952a, p. 462). Logical empiricists who embrace DN explanation and HD confirmation will likely latch on to HH's use of the word "definitions." Yet, HH do not mean by this term what some philosophers might mean, namely, some stipulation about how they will use notation. Instead, HH are advancing an empirical hypothesis about the relations among "real properties" of the membrane. More specifically, they have in mind the hypothesis that these quantities are linearly related. Although it does not appear in Hodgkin and Huxley (1952a), their idea is that the quantities are related as specified by a rearrangement of Ohm's law, $V = IR$, where the voltage is given by $E - E_{Na}$ and g is the inverse of R. This would have been obvious to HH's fellow physiologists but was only explicitly mentioned in the later papers. The idea that they are advancing an empirical hypothesis, rather than a stipulative definition, is important insofar as it helps philosophers of science resist the empiricist interpretations of HH's work.

On the scientific front, one can foresee in these equations their general strategy for getting an experimental handle on conductances. E is the voltage set by the voltage clamp. Following their results of the third experiment in Hodgkin and Huxley (1952a), they had shown how to calculate E_{Na} and E_K. Finally, given their resolution of the ionic current into sodium and potassium currents, they could measure the time course of I_{Na} and I_K. Thus, an important part of the work following the first three experiments was to show how voltage clamp results could provide for an indirect means of ascertaining the conductances. This was a significant theoretical and empirical step forward that was explored in the third and fourth papers of the 1952 series. Roughly speaking, the third paper concerns the "activation" of membrane permeability, whereas the fourth paper concerns the "inactivation."[45]

Why was this empirically and theoretically important? Recall that, in explaining the action potential, Bernstein's membrane hypothesis posited changes in membrane permeability. Bernstein, however, had no means of isolating changes in membrane permeability from changes in the potential. Having established that sodium permeability rises and falls in the early phase of the action potential and that potassium permeability rises with a delay and remains stable, HH have apparently made the scientific problem

[45] For this terminology, see Hodgkin and Huxley (1952c, p. 497).

more difficult. There is not merely ionic permeability to consider. There are two ion permeabilities with their distinct time courses to consider. How can a scientist get an empirical handle on these permeability changes? By postulating a relationship between the time course of conductances and the time courses of ion currents, membrane potentials, and ionic potentials. Such features of conductances began to pave the way for a compositional explanation of instances of the conductance of the membrane in terms of activity instances of individuals in the membrane.

It is worth emphasizing this theoretical strategy since little in HH's discussion draws attention to its significance. Their paper has no historical contextualizing. There is only a minor section break. Moreover, philosophers of science who have written about the Hodgkin–Huxley model have often noted that HH could not provide a compositional explanation of the changes in membrane permeability.[46] While this last point is true, it understates the importance of HH's experimental and theoretical advance. Further, it ignores just how early in the course of scientific research it would have been had HH come up with the correct account of the "physical basis" of membrane permeability.

5.4.2 "The Components of Membrane Conductance"

Although there is much of philosophical importance in Hodgkin and Huxley (1952b, 1952c), the present account will be limited to some of the first experiments on sodium conductance in "The Components of Membrane Conductance." These experiments bear out the role of compositional abduction in HH's interpretation of their experimental results. The overarching experimental method was to voltage clamp axons at different potentials, recording the currents at the times of change. In effect, the different potentials immediately changed the driving force for the current, but the changes in membrane permeability followed with some delay. As HH put it, there was a discontinuity in the currents, but continuity in the conductances. This temporal difference allowed HH to infer the time course of conductance changes.

In Figure 5.3, the left column represents the time course of command voltages, whereas the right column represents the time course of currents. In the top row, a single depolarization of 41 mV (represented in a on the left) gives rise to the usual initial rise in inward current eventually declining and giving way to an outward current (represented in A on the right). HH assume without comment that there is an initial capacity current at $t = 0$.

[46] See, for example, Craver (2008) and Levy (2014).

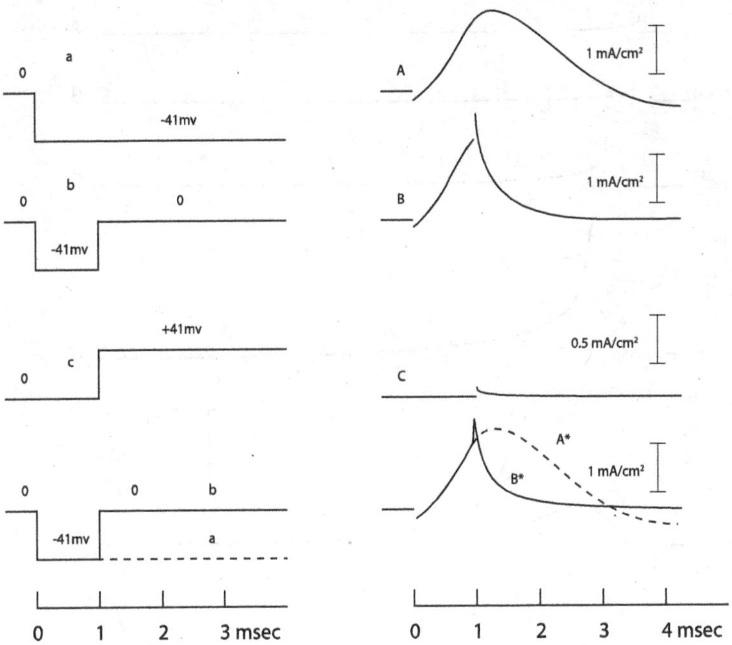

Figure 5.3 Left-hand column: a, b, c, time course of potential difference between external and internal electrode. Right-hand column: A, B, C, records of membrane current associated with changes in membrane potential shown in left-hand column. (The amplification in C was 90% greater than that in A and B.) A*, B*, time course of ionic currents obtained by subtracting capacity current in C from A and B. Axon 25; temperature 5°C. Redrawn from Hodgkin and Huxley (1952b, p. 474, figure 1).

In the second row, a pulse of depolarization by 41 mV is followed by a restoration of the potential to the resting potential after 1 msec (represented in b on the left). The current initially rises just as before for the first msec, but there is a jump in the value of the inward current immediately after the restoration of the resting potential (represented in B on the right). HH assumed that two things happen at t = 1 msec. First, there is a jump or discontinuity in inward current, because of the jump or discontinuity in the driving force on sodium ions and that there is an inward capacity current. The experiment depicted in the third row explores this. In c, there is a single hyperpolarization step that induces a current spike shown in C which HH abductively infer is a capacity current. What would explain this short-lived current in the direction of the potential change? A capacity current. They then subtract this capacity current from A and B to get the reconstructed curves in A* and B* in the fourth row.

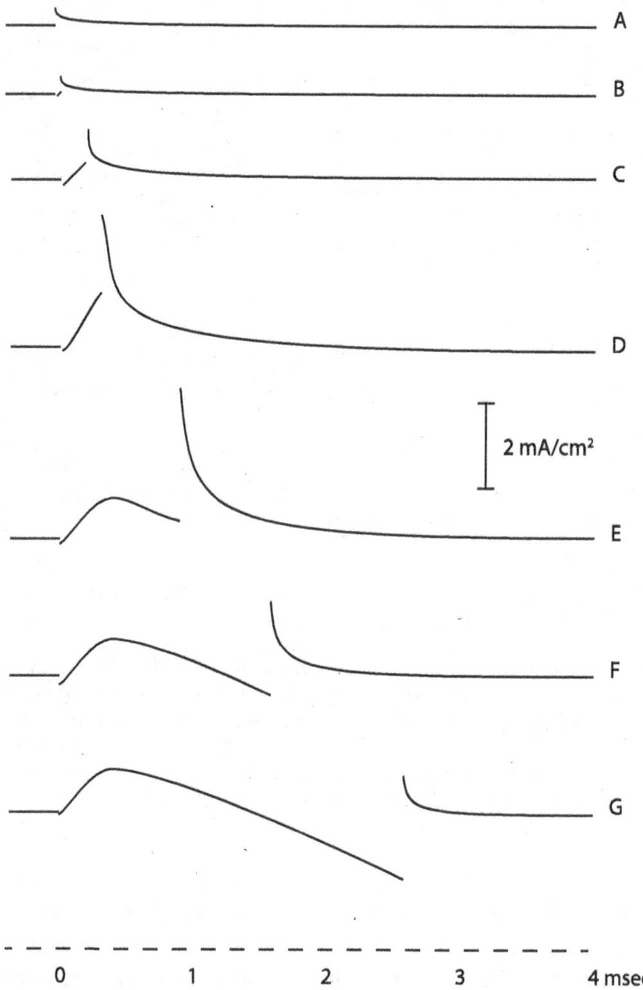

Figure 5.4 Records of membrane current associated with depolarization of 97.5 mV lasting, 0.05 (A), 0.08 (B), 0.19 (C), 0.32 (D), 0.91 (E), 1.6 (F) and 2.6 (G) msec. Redrawn from Hodgkin and Huxley (1952b, p. 475. figure 2).

In a second experiment designed to reveal the discontinuity in the inward current, HH set the initial pulse to a value close to the sodium potential. In this case, the driving force during the pulse, hence the current during the pulse, would be near zero. This follows from the assumption that $I_{Na} = g_{Na}(E - E_{Na})$. HH then varied the duration of the pulse. In curves A–D of Figure 5.4 above, the discontinuity grows. Why? HH

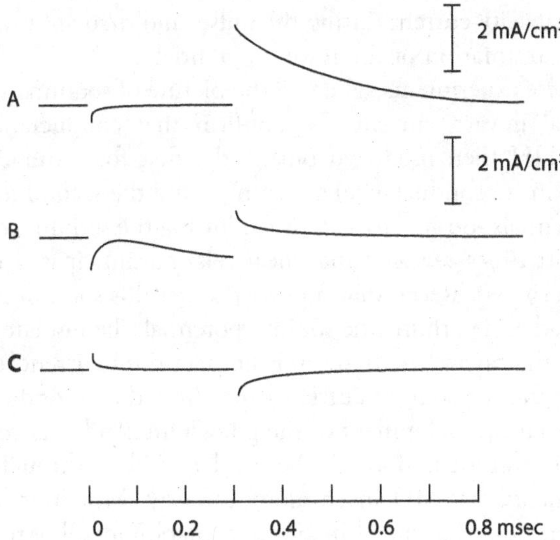

Figure 5.5 A, membrane current associated with depolarization of 110 mV lasting 0.28 msec; nerve in seawater. B, same, but with nerve in choline seawater. C, membrane currents associated with an increase of 110 mV in membrane potential; nerve in choline seawater. Axon 25; temperature 5°C. Redrawn from (Hodgkin & Huxley, 1952b, p. 476. figure 3).

reason that the depolarization triggers increasing membrane permeability that peaks at about 0.5 msec. Because the membrane permeability increases, the current increases. In E–G, however, the inward current declines as potassium permeability increases and potassium ions begin to leave the axon.

In a third experiment designed to nail down the role of sodium, HH began with a short 0.3 msec pulse depolarizing the cell to the sodium potential, then restoring the resting potential (see Figure 5.5). The pulse produced a capacity current but no ionic current. This was expected as there was no driving force on the sodium, because the cell was depolarized to the sodium potential. After the pulse, however, the rest potential gave rise to an inward driving force. This result is shown in A. In the next condition, the axon was placed in choline seawater. Here, the results are shown in B. During the pulse, some of the intracellular sodium flowed down the concentration gradient generating an outward current. Upon restoration of the resting potential, there was no extracellular sodium to flow into the axon, hence no inward current. Finally, in C, HH show the results of an increase in polarization of 110 mV, followed by a restoration of the rest potential in a sodium-containing medium. This generated only an outward capacity current during the pulse, and then an inward capacity current upon restoration of the resting potential.

Each of these experiments validated the picture of sodium as responsible for the initial inward current. To confirm that conductances satisfied Ohm's law, HH then used two pulses, the first for a duration of 1.53 msec (to maximize conductance) at -29 mV, and the second for a range of different potentials -60 mV to +30 mV. Plotting the sodium conductances for these short times revealed the linear relation anticipated in Hodgkin and Huxley (1952a). Recall that, in order to get this sodium conductance curve, HH had to determine the sodium potential, then isolate the sodium current from the capacity current and the potassium current, then experimentally separate the sodium current from the sodium conductance. And there was a parallel development of the potassium conductance curve. The thread of this argumentation can be hard to follow through the many careful comments that HH insert along the way. Moreover, it is easy to lose sight of the idea that the ultimate confirmation for all of this apparatus is that it enables HH to explain the multiple sets of experimental results they obtained along the way.

HH's development of the sodium and potassium conductance curves was an important experimental achievement. It set the stage for the advancement of a compositional explanation of the conductances. In point of logic, over the course of Hodgkin and Huxley (1952a, 1952b, 1952c), HH might have been able to determine enough features of the sodium and potassium conductances to enable them to abductively confirm some hypothesis regarding the individuals in the membrane whose activity instances might explain one or another experimental result.[47] Their work might have done the equivalent of measuring enough features regarding rat maze navigation to enable the psychologist to infer that the rat was navigating using a cognitive map, rather than a local response strategy. In the end, however, the best HH could do was to disconfirm a membrane permeability hypothesis they had entertained when they first conceived their experiments.[48] On this rejected view, negatively charged "carrier" molecules shuttle the positively charged ions across the membrane in response to depolarization. The fatal flaw in this model was that it predicts an initial *outward* negative current that was not detected. In other words, the hypothesis led to a conclusion about

[47] Hodgkin and Huxley (1952b, pp. 495–496) and Hodgkin and Huxley (1952c, pp. 505–506) provide concise summaries of potential explananda.

[48] Hodgkin and Huxley (1952d, pp. 502–504). See also Hodgkin (1976, p. 15).

something that was not detected, something that should be interpreted as an instance of compositional abductive disconfirmation.

5.5 Summary

The principal aim of the present chapter has been to illustrate how the philosophical account of singular compositional abduction developed in Chapters 2–4 characterizes some of Hodgkin, Huxley, Curtis, and Cole's reasoning about their experimental results. It is meant to illustrate how the account of singular compositional abduction makes explicit so much of what is merely implicit in this body of mid-twentieth-century physiological research.

As scientifically important as the HH theory is and as familiar as it is in the context of the New Mechanism, historians and philosophers of science have not even scratched the surface of the complexity of this work. Few of HH's papers are even cited in the philosophical literature. Unfortunately, the present chapter does not go far in rectifying the situation. Far from being a complete or even thorough examination of the mid-twentieth-century work on the action potential, it even ignores large parts of HH's 1952 papers. There is simply too much ground to cover the subtleties of the experiments and the theoretical background to them. This chapter only touches on one aspect of the mid-twentieth-century neuron physiology, namely, some instances of the use of singular compositional abduction.

CHAPTER 6

The Hermann Grid Illusion

The history of the development of the Hodgkin–Huxley theory of the action potential provides one detailed illustration of singular compositional abduction. This supports the claim that there exist scientific uses for such abductive inference. The philosophical importance of singular compositional abduction will, however, be enhanced insofar as there are grounds for thinking its use extends beyond this case. One basis has been the brief review in Chapter 4 of intralevel experiments that *prima facie* appeal to compositional abduction. Another basis provided here is an additional case study somewhat removed from the study of the action potential.

Central to this next case study are many intralevel experiments in which scientists present a participant with some visual stimulus and then measure that participant's response. The individual responses constitute data that are aimed at results regarding what the participant perceives. These are psychophysical experiments in which scientists manipulate a participant in an experiment and measure a response, then hypothesize some perceptual activity. The perceptual results – the instances of perceptual activities – are then interpreted in terms of entities whose activity instances are hypothesized to compositionally produce visual perceptions.

Recognizing the role of psychophysical experiments in the study of the Hermann grid illusion suggests that compositional abduction may also play a role in psychophysical experiments generally. In Chapter 1, I noted one problem with the use of case studies: How can one generalize from one or a few cases? As an answer, I proposed that one refer to features of the case that are to be found in other cases. Here I have indicated what I take to be a feature of the study of the Hermann grid illusion that may support its generalization to other cases.

Section 6.1 presents the initial description and explanation of the Hermann grid illusion in Hermann (1870). Section 6.2 then turns to the standard textbook account of the illusion in terms of RGCs introduced in Chapter 4. Section 6.3 describes some experiments abductively

Figure 6.1 Simultaneous contrast. The medium gray boxes on the left and the right are physically the same but are perceived differently due to the surrounding context.

supporting the RGC hypothesis, whereas Section 6.4 describes some that challenge it. The goal of this case study is not to show that every experiment that is relevant to the grid illusion is an instance of singular compositional abduction. Obviously, the study will only be a partial review of the literature that has touched upon the illusion. The modest goal here is to make a case that singular compositional abduction has had a significant role in the scientific study of the Hermann grid illusion. Indeed, Section 6.5 argues that not all instances of scientific hypothetical reasoning about experimental results directed toward compositional hypotheses should be construed as compositional abduction. It will broach the idea of singular interlevel abduction.

6.1 Ludimar Hermann's Report of the Grid Illusion

Recall from Chapter 4 that, in one version of the Hermann grid illusion, there are intersecting vertical and horizontal white bars on a black field wherein scintillating gray smudges appear at the intersections peripheral to the viewer's fixation point (see Figure 4.3, Chapter 4). In his discussion, Hermann supposed that there are two features of the illusion that are to be explained: (1) there are smudges at the crossing points of the white bars and (2) the smudges appear only in peripheral vision.

Hermann's explanation of the illusion involved two hypotheses: there are retinal elements that produce simultaneous contrast and these retinal elements are sensitive to spatially limited regions. Simultaneous contrast refers to the fact that the perception of the lightness of one region is influenced by the lightness of an adjacent region. For example, in Figure 6.1, a medium gray box of a given luminance is perceived to be darker when surrounded by a very light gray region (on the right) than when surrounded by a very dark gray region (on the left). All this despite the two medium gray boxes being qualitatively the same. The second

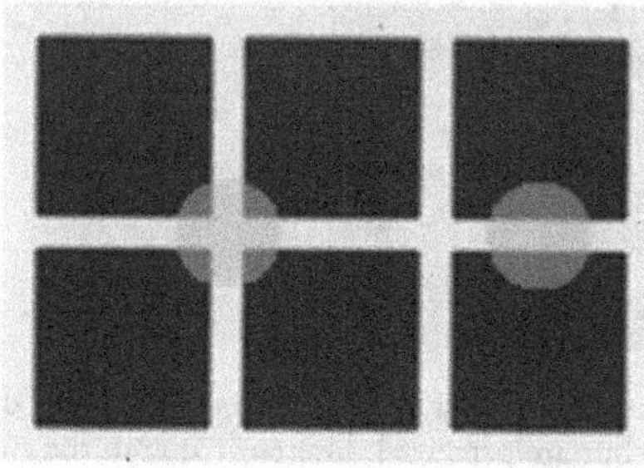

Figure 6.2 Hermann's explanation of the smudges at crossing points: There is more black in the circular region on the right than in the circular region on the left, so the region on the right has more brightening by simultaneous contrast than does the region on the left.

hypothesis is that the retinal elements that influence the perception of a given region are limited in size.[1] Thus, in Figure 6.1, the light gray field on the right has a greater influence on the right-hand medium gray box than it does on the left-hand medium gray box and the very dark gray field on the left has more influence on the left-hand medium gray box than it does on the right-hand medium gray box.

With this background, Hermann's explanation of the smudges at the crossing points of the white bars is clear:

> The apparent brightness of each point on the white grid depends on the amount of black which exists in a certain area around it. If one assumes the diameter of this area to be larger than the width of the white stripes, then each point on the intersections receives in its surround less black than any other point on the white stripes; its brightness will thus be less enhanced by contrast and must therefore appear darker. (Hermann, 1870, p. 14, translation from Spillmann, 1994, p. 692)[2]

Figure 6.2 illustrates Hermann's proposal.[3]

The second feature of the illusion is that the smudges do not appear at the viewer's fixation point. If the viewer changes her point of fixation, the

[1] Contemporary discussions might frame Hermann's idea in terms of receptive field sizes. This conceptualization would be anachronistic. Sherrington (1906) is widely credited with introducing the concept of a receptive field size as referring to the area of skin that could be stimulated to elicit a scratch response. Hartline (1938) extended the concept to that area on the retina from which a response in another cell could be elicited.

[2] Note Spillmann's assumption that Hermann's "Die Erklärung" is appropriately translated as "an explanation."

[3] Note that on Hermann's account, unlike more recent accounts, the illusion is due to excess brightening of the bars away from the crossing points, rather than a darkening of the region of the crossing points. Cf., Spillmann (1994, p. 692).

Figure 6.3 In central vision, the relevant parts of the visual system are sensitive to smaller regions so that they do not differ in their generation of simultaneous contrast.

region where the smudges appear changes. In Hermann's account, the smudges do not appear at fixed crossing points. Hermann's explanation of the absence of the illusory perception in central vision, illustrated in Figure 6.3, is that the central regions responsible for simultaneous contrast are smaller than the peripheral regions for simultaneous contrast:

> Furthermore, that darkening does not occur in fixed crossing points is explained by the assumption that, for objects that are imaged on the center of the retina, the perimeter within which the simultaneous contrast applies is smaller than for other areas. (Hermann, 1870, p. 14)

Thus, Hermann postulates individuals and activity instances that compositionally explain the illusion.

Before moving on to subsequent historical developments, let me review how various features of the theoretical framework developed in Chapters 2 through 4 describe what Hermann was doing. To begin with, a specific human's perceiving the illusion at a given time is an activity instance of an individual. Hermann believes this is explained by unspecified activity instances of unspecified retinal elements. More concretely, Hermann does not articulate the explanation in terms of, say, firings of action potentials in retinal ganglion cells (RGCs) or graded potentials in photoreceptors. This is an instance of an "individual-free" and "activity-free" explanans.

A second thing to note is the enabling role of the area to which a retinal element responds. Size figures into the explanation of both of the explananda Hermann tries to explain. Regarding the first explanandum, the size of the unspecified retinal elements – a property instance – enables it to engage in an unspecified activity instance that implements an instance of a perceived smudge when the element falls at a crossing point. By contrast, that size instance does not enable an instance of that unspecified activity when the retinal element falls on a horizontal or vertical bar. Regarding the second explanandum, the small size of the unspecified retinal element in central vision explains why there are no illusory smudges at the fixation point, but only in peripheral vision. So, an accurate recounting of Hermann's explanations must recognize both the activity and property instances.

Note, as a third thing, that Hermann's experiment involved presenting grids of various descriptions to humans and soliciting their verbal responses. In these experiments, there is what subjects perceive and there is what subjects say they perceive. Hermann assumes that these verbal reports often or typically indicate what subjects perceive. To use the terminology of Chapter 3, the data in this case are the human responses and the results are the human perceptions. In more recent psychological work, there is often considerable debate over the extent to which verbal reports accurately indicate what subjects perceive. There is a worry that what a subject reports is influenced not only by what the subject perceives, but also by post-perceptual processes of judgment.[4] Hermann, however, did not have those concerns.

Fourth, note that Hermann used intralevel psychophysical experiments wherein he presented a subject with a visual stimulus, then took the subject's verbal report. Based on these experiments, he postulated that unspecified individuals in the brain generate simultaneous contrast. As reviewed in Chapter 5, Hodgkin and Huxley also used intralevel experiments – manipulating axonal membrane potentials and measuring subsequent currents – as a basis for drawing inferences about ion fluxes and conductances. Hodgkin and Huxley could specify the explanans entities and activities. Hermann's inferences, however, left hypothetical entities – individuals and their activity instances – unspecified. Of course, while some of Hodgkin and Huxley's inferences did specify the hypothetical entities engaged in activity instances, that is, movements of sodium and potassium ions, some of their other inferences did not, that is, what individuals in the membrane change during changes in membrane permeability.

Finally, note the retinal activities to which Hermann alludes are not the biological bases of the *experiences* of the illusion. Those biological bases, whatever they are, are widely assumed to be cortical. Thus, I propose that Hermann thinks that the retinal activity instances compositionally explain a contributing activity to the perception of the illusion, where the

[4] The literature on this topic is so extensive that there is little point in trying to refer to any significant number of works. I will, however, refer to just one paper that is especially important: Firestone and Scholl (2014).

perception of an illusion refers to a process that begins in the retina and ends somewhere in the cortex.

6.2 Baumgartner's Retinal Ganglion Cell Theory

Developments in the 1950s helped specify Hermann's physiological hypotheses.[5] Barlow (1953) reported the discovery of ganglion cells in the frog retina that displayed lateral inhibition.[6] In the first step in one of the relevant experiments, Barlow isolated cells that respond either to the onset or offset of a spot of light in a cell's receptive field. In the second step, Barlow showed that adding a second spot of light a small distance from the first would inhibit responses to both the onset and offset of the first spot. In the discussion of his results, Barlow proposed that, "Simultaneous contrast effects are presumably caused by an inhibitory mechanism similar to the one described; a white spot surrounded by black looks brighter (more impulses) than a white spot of the same intensity surrounded by grey, because the grey falls on the inhibitory fringe of the ganglion cells, and so inhibits their discharge (causes fewer impulses)" (Barlow, 1953, p. 85). Unlike Hermann, Barlow tied simultaneous contrast to specific activity instances, namely, electrical discharges, and to specific retina elements, namely, ganglion cells.

Kuffler (1953) provided important complementary results in experiments with cats. Simplifying somewhat, Kuffler found that some ganglion cells responded to spots of light in their receptive field centers, but were inhibited by spots of light in a surrounding annulus. These were ON-center, OFF-surround cells. In addition, he found cells with a complementary organization. The OFF-center, ON-surround cells were inhibited by spots of light in their receptive field centers but were responsive to spots of light shown in a surrounding annulus. Like Barlow, Kuffler drew attention to part of the perceptual significance of his results: "This should be advantageous in the perception of contrast" (Kuffler, 1953, p. 65). Further, he was specific about the activity instances and individuals.

In his three-paragraph paper of 1960, Baumgartner revised Hermann's account, incorporating Barlow and Kuffler's physiological work.[7] In outline, he proposed that RGCs were the retinal elements to which Hermann referred, that the firings of RGCs were the activity instances implementing a perception of the illusion, and that the sizes and response

5 For further historical discussion, see Spillmann (2014).

6 Later discussions describe these as "retinal ganglion cells."

7 Baumgartner (1960).

properties of RGCs were the enabling property instances. In other words, Baumgartner realized that RGCs and their activity and property instances filled out the details of the explanation developed by Hermann.

For the case of white bars on a black background, Baumgartner hypothesized that when an ON-center/OFF-surround cell falls at an intersection, there is more OFF-surround stimulation, thereby making for a weaker cell response and a smudge. By contrast, when an ON-center/OFF-surround cell falls along a line away from an intersection, there is less OFF-surround stimulation, thereby making for a stronger cell response and no smudge. For the case of black bars on a white background, there is a correlative story implicating OFF-center/ON-surround cells.

It is worth noting that my interpretation is based on the following translation from Baumgartner's German:

> The darkening of the light bars at the crossings is then explained by the here more pronounced edge zone inhibition, which reduces the activation of the on-neurons in comparison to the other bar sections. Conversely, the lightening of the dark bars in the crossing region is to be understood by the lower activation of the off-neurons from the edge zone since the positive dark impression correlates with the frequency of the dark-activated neurons. (Baumgartner, 1960, p. 22)[8]

In adopting this translation, I assume that the German "erklärt" is correctly translated as "is explained by" and that "verstehen" is correctly translated as "is to be understood." Further, I assume that Baumgartner's choice between "erklärt" and "verstehen," although they differ in meaning, does not reflect any serious scientific difference. Instead, they are for Baumgartner's purposes, merely different stylistic choices. I do not presuppose that these interpretations are completely obvious, but they are plausible.

Like Hermann, Baumgartner supposes that he can explain why the smudges do not appear in central vision in terms of the sizes of the receptive fields of retinal ganglion cells. Baumgartner's explanation is essentially the same as Hermann's explanation, illustrated in Figure 6.2. The additional ON-center/OFF-surround structure of RGCs leaves Hermann's account unchanged: both the ON-center and OFF-surround of RGCs in central vision fall within the white crossing point, so there is no OFF-surround contribution to darkening.

One of the philosophically interesting features of Baumgartner's discussion lies in his final paragraph. He reasons that, if the illusion is strongest

[8] Many thanks to Beate Krickel for reviewing this translation.

when the size of the ON-center matches the angular width of the intersecting bars, then this provides an estimate of the ON-center diameter of >25 μ. What is interesting is that this implies that Baumgartner thought that RGCs produced the illusory perception. If RGCs did not generate the phenomenon, then why would the viewing of the grid be relevant to RGC receptive field size? Although one might reasonably conjecture that Hermann thought that the individual-free explanans he proposed in 1870 was correct, there is no firm textual basis. However, Baumgartner adds some brief comments that commit him to RGCs being involved in the Hermann grid illusion. That further implies that he thinks that singular compositional abduction provides evidence supporting the RGC hypothesis.

6.3 Development of the Retinal Ganglion Cell Theory

Baumgartner's paper did not report on experiments he performed, so he was not strictly speaking interpreting his experimental work. Spillmann (1971), however, treats the RGC hypothesis in a properly experimental context. For the sake of brevity, I will focus on Spillmann's discussion of ON-center/OFF-surround cells, and his experiments to estimate their receptive field sizes. As with Baumgartner, the method of estimating sizes presupposes that, in fact, retinal ganglion cells are part of the neurophysiological basis of the illusory perception.

In brief, Spillmann hypothesized that a cell will produce a smudge of maximal intensity if it is precisely sized and centered at a standard crossing point on the grid. This is Baumgartner's assumption in the final paragraph of his paper. It is the configuration of the second cell from the right in Figure 6.4 below. The two cells on the right would have this same maximal intensity, since the portion of the bars outside of the cell's receptive field would not influence its firing. Decreasing the size of the bars outside the ON-center and detecting just noticeable differences in the intensity of the illusion would, therefore, provide an estimate of the entire receptive field. In contrast to this, the two cells on the left-hand side of the figure would produce weaker smudges, because they have less OFF-surround stimulation. Of the two cells on the left, the cell farthest to the left would produce a weaker smudge than the cell second from the left. All of this was hypothetical reasoning.

Spillmann checked the consequences of his hypotheses in his "Series A" experiments. In these experiments, subjects first view the standard Hermann grid to learn to recognize the illusion. Then, modified grids, such as the one in Figure 6.5, were presented to them. Eight standard

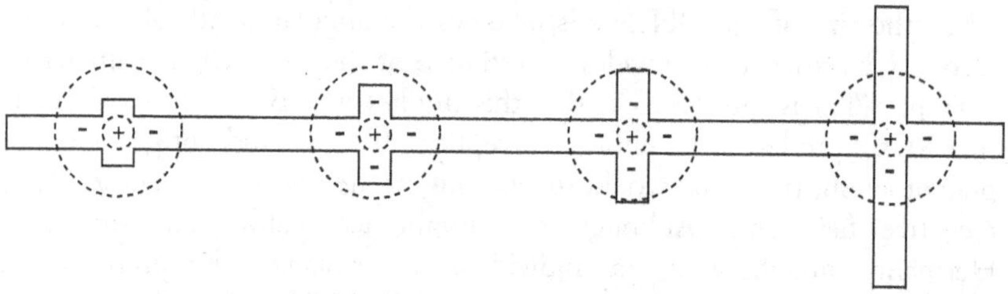

Figure 6.4 Illustration of Spillmann's hypothetical thinking regarding the Hermann grid illusion. A RGC on the far right would produce the maximal amount of darkening. So, would the RGC second from the right since the portion of the vertical bar that falls outside the cell's receptive field would not influence its response. The two RGCs on the left, by contrast, would produce submaximal responses, with the cell on the far left producing a weaker response than the cell second from the left. Redrawn from Spillmann (1971, p. 284, figure 2b).

intersections were along the sides, whereas the central regions had eight experimental bars symmetrically shortened to various lengths. (Note that one of the experimental bars was a "null bar," with no vertical length. This is in the upper middle left of Figure 6.5.) Subjects were first asked to fixate experimental bars to see if a smudge was present.[9] If so, then subjects were asked to compare the experimental bar with a nearby standard intersection. The comparisons were either "weaker," "equal," or "stronger." Seven college undergraduates with 20/20 uncorrected vision or better, with no significant astigmatism, were participants.

Results from the Series A experiment are shown in Figure 6.6. The leftmost curve with squares plots the probability of a subject reporting no effect (the ordinate) versus the length of the experimental bar in arcmin (the abscissa). Experimental bars shorter than about 11 arcmin have less than a 50% chance of inducing a reported effect. The middle curve with circles plots the probability of a subject reporting an experimental effect that is weaker than the standard effect. The right curve plots the probability of a subject reporting an experimental effect that is the same as the standard effect. Thus, experimental bars longer than about 19 arcmin are more than 50% likely to be indistinguishable from the standard bars. No experimental bars were reported to produce a stronger effect than a standard bar. There were no differences between horizontal experimental bars and vertical experimental bars. Spillmann interprets the 19 arcmin

[9] It is unclear from Spillmann's discussion how fixating an intersection was consistent with the illusion standardly appearing in the visual periphery.

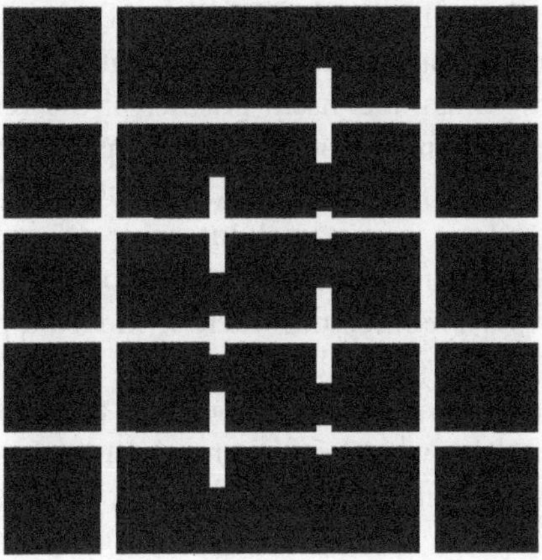

Figure 6.5 Stimuli from Spillmann's "Series A" experiments. Eight "standard" intersections are on the right and left columns. "Experimental" intersections with shortened bars are shown in the middle. Subjects compared "experimental" intersections with their nearest "standard" intersections. Another image using horizontal bars was also used. Redrawn from Spillmann (1971, p. 284, figure 2a).

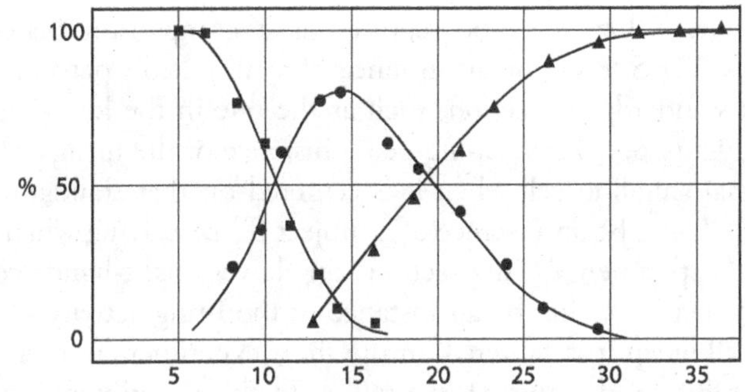

Figure 6.6 Percentage distribution of responses for series A as a function of bar length (in min of arc). The first intersection of the curves at the 50% level (squares and circles) indicates the minimal angular length of a pair of bar extensions necessary to produce a just noticeable darkening effect. The second intersection at the 50% level (circles and triangles) marks the threshold at which the grid illusion reaches a maximum. This length is considered equivalent to the diameter of a foveal perceptive field. Redrawn from Spillman (1971, p. 286. figure 3).

figure as an estimate of the receptive field size, matching the hypothetical picture of the right half of Figure 6.4.

Notice, to begin with, the overarching structure of Spillmann's argumentation. He begins with the hypothesis that the biological basis of the grid illusion lies in the activity of RGCs. Using Figure 6.4, he reasons that if this is so, then as vertical bars increase in length there will be a point at which darkening first manifests itself. Further, if the vertical bars increase in length even more, then they will increase up to a value at which the vertical bars reach the full extent of the ON-center, OFF-surround receptive field and, therefore, a maximal intensity of the illusory smudge. He then compares the consequences of this hypothesis against an experimental determination. He finds that there is indeed a point at which the increasing length of the vertical bars leads to a maximal darkening. The dependencies outlined by the RGC hypothesis are found to obtain. The RGC hypothesis explains why there is a maximum bar length and thereby enables Spillmann to estimate the total receptive field to be about 19 arcmin. Each measurement corresponds to an instance of a singular compositional abductive inference. Here again is an instance of hypothetical reasoning that I take to be better understood as relying on abductive confirmation than HD confirmation.

Next, consider the structure of the experiment more closely. Each data point represents a judgment of "weaker," "equal," or "stronger." (Again, in the actual experiment, no judgments were "stronger.") Each datum is a verbal report. What is to be reported is a comparison between two perceptions. Let S Ψ-ing be an instance of a subject S's perception when S fixates a standard intersection, such as the one in the lower right-hand corner of Figure 6.5. Let x_i ϕ_i-ing be an instance of the firing activity of a single retinal ganglion cell when it is centered on that standard intersection. Let S Ψ*-ing be an instance of a subject S's perceiving when S fixates the short experimental intersection on lower right-hand corner of Figure 6.5. Let x_i ϕ_i*-ing be an instance of the firing activity of a retinal ganglion cell when it is centered on the short experimental intersection in the lower right-hand corner of Figure 6.5. In this experiment, an individual S is asked to make multiple comparisons of S Ψ-ing versus S Ψ*-ing.[10]

Notice, in addition, that each datum – each judgment of "weaker," "equal," or "stronger" – is a spatiotemporal particular based on a data target – an introspective comparison of two further spatiotemporal particulars, S Ψ-ing and S Ψ*-ing. S Ψ-ing and S Ψ*-ing are distinct activity instances of S; they are distinct results. One is, as it might be, an instance

[10] Note that, in this experiment, Spillmann does not have an explanation of S Ψ-ing in terms of x_i ϕ_i-ing or of S Ψ*-ing in terms of x_i ϕ_i*-ing, but only of S Ψ-ing differing from S Ψ*-ing in terms of x_i ϕ_i-ing and x_i ϕ_i*-ing differing. These explanations are distinct from other controlled experiment explanations, as discussed in Chapter 3.

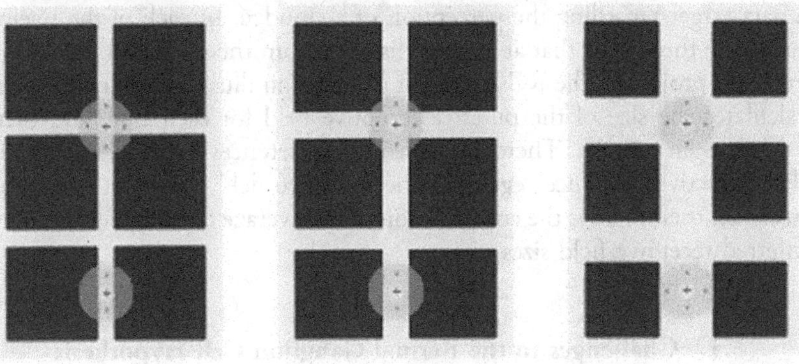

Figure 6.7 The Hermann grid effect weakens and finally disappears when the retinal image of the intersection is either too small or too large with respect to the perceptive field center. The reduction of the effect with greater than optimal bar widths is attributed to the decreasing difference in lateral inhibition between perceptive field centers illuminated by bar or intersection, respectively.

of a perception of a dark smudge, whereas the other is an instance of the perception of a light smudge. Further, Spillmann assumes that S Ψ-ing is singularly compositionally explained by x_i ϕ_i-ing, whereas S Ψ^*-ing is singularly compositionally explained by x_i ϕ_i^*-ing. Spillmann can, of course, generalize from these instances, concluding that the perception of smudges is implemented by the firings of RGCs, but those are generalizations from the singular compositional explanations.

Consider now a second experiment that again illustrates a role for singular compositional abductive inference. Spillmann reasons that, if the grid illusion is due to RGCs, then there should be a bar width at which the smudges are most intense with the intensity decreasing at greater bar widths (see Figure 6.7). To test this, he had participants view his "Series A" grids moving the participants nearer or farther from the grids while seated in a roller chair. For six student subjects, there was an ascending and a descending threshold (n = 12). Pooled and averaged, the results yield an average value of 18.7 arcmin for the entire receptive field. As predicted by the RGC theory, there was indeed a maximum bar size for the illusion.

Notice that, in this argumentation, there is hypothetical reasoning and it is used to confirm the RGC hypothesis. If the grid illusion were due to RGCs, then there should be a maximal bar size. There is a maximal bar size, which is explained by the RGCs generating the illusion. Further, each of the twelve subject judgments – each verbal report – is a datum aimed at a data target regarding the perception of a smudge. In each of the twelve cases, the thought is that at a given distance from the screen on which the grids are projected, the RGCs do not generate an illusion. Spillmann then calculates the size of the putative receptive field for each case, and then calculates an average. There are, thus, two inferences here. First, there is the abductive inference regarding the receptive field size for a particular instance, then there is the computation of the average from the abductively inferred receptive field sizes.

6.4 Challenges to the Retinal Ganglion Cell Hypothesis

In this section, I will review five experiments that I take to illustrate scientists using compositional abduction in attempts to disconfirm the RGC hypothesis. Two are from Wolfe (1984) and three are from Schiller and Carvey (2005). I have selected these arguments because they are relatively accessible. Further, I have chosen arguments from different papers to show that using compositional abduction for disconfirmation in experimental contexts is not idiosyncratic to one scientist.[11] I think the picture of singular compositional abduction from Chapter 4 fits these examples.

Wolfe (1984) provides two simple experiments illustrating abductive disconfirmation of the RGC hypothesis. He notes that the RGC hypothesis focuses exclusively on local factors, roughly speaking, what falls within the scope of a single receptive field. He reasons that, were this true, then the global structure of the display should not influence the perception of the illusion. The disconfirmation arises when the global structure of the display is found to influence the perception.

For the first experiment, Wolfe reasons that "If the Hermann grid illusion could be explained entirely on the basis of the local activity of concentric receptive fields, then the effect should be equally strong with one or many intersections" (Wolfe, 1984, p. 34). Clearly, Wolfe is reasoning hypothetically. Therefore, the experiment is intended to show

[11] There are other examples of compositional abductive disconfirmation in Schiller and Carvey (2005). Further, Spillmann (1971, pp. 294–295) uses compositional abduction to try to disconfirm the hypothesis that eye movements implement the grid illusion. Schiller and Carvey (2005, p. 1386) uses compositional abduction to disconfirm the hypothesis that the grid illusion is implemented by rod/cone interactions. In the interests of full disclosure, it is unclear to me what philosophical theory fits Schiller and Carvey's seventh argument against the RGC hypothesis. Their argument is that the spatial arrangement of retinal ganglion cell receptive fields is not what has been assumed by the RGC theory. See Schiller and Carvey (2005, p. 1382).

Table 6.1 *Parameters of the thirteen grids in Wolfe's first experiment*

Grid	Number of intersections	Length of block (deg)	Length of grid (deg)
1	1	1.6	4.0
2	4	1.6	6.0
3	9	1.6	8.0
4	16	1.6	9.8
5	25	1.6	11.6
6	36	1.6	13.5
7	1	1.2	3.4
8	9	1.2	6.0
9	36	1.2	11.0
10	1	2.3	5.6
11	9	2.3	10.7
12	1	3.6	8.0
13	4	3.6	11.6

that the effect is not equally strong with one or many intersections. To do this, Wolfe used grids that differ in the number of intersections, the size of the blocks, and the overall length of the grid, as shown in Table 6.1. Wolfe also used both black-on-white grids and white-on-black grids. The ten subjects in the experiment were taught a rating scale of 1–9, wherein 7 was indexed to a grid with 36 intersections (grid 6 in Table 6.1). 1 represents no illusion and 8–9 represent an illusion stronger than the reference grid 7. Wolfe found a strong effect of number of intersections.

For the second experiment, Wolfe hypothetically reasoned, "If each intersection produces the same illusion regardless of its surround, then 25 irregularly placed intersections should produce the same magnitude of illusion as a 5 x 5 square grid" (Wolfe, 1984, p. 37). Wolfe disconfirmed the RGC hypothesis by showing that stimuli consisting of irregularly placed intersections – essentially fields of randomly placed "+"s – produced a weaker illusion than do stimuli consisting of standard grids. For this study, only black "+"s on a white field were tested. Stimuli with 1, 4, 9, 16, and 25 intersections arranged irregularly were compared with standard grids having 1, 4, 9, 16, 25, and 36 intersections. The results represented in Figure 6.8 show that the regular organization of the standard grid stimuli has an effect, but also that there is a weak effect due to the number of intersections. (Wolfe does not provide a statistical analysis.)

Schiller and Carvey (2005) provides several reasons that they take to challenge the hypothesis that the perception of the Hermann grid illusion

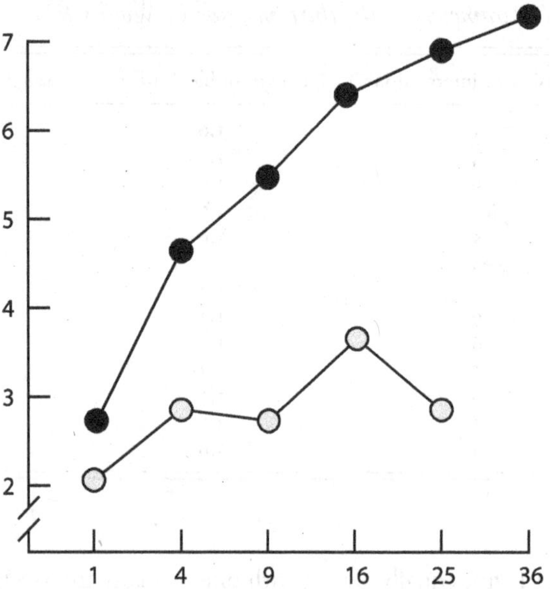

Figure 6.8 The magnitude of the illusion (abscissa) as a function of the number of regularly and irregularly placed intersections (ordintate). Redrawn from Wolfe (1984, p. 38, figure 5).

is generated by RGCs, but support the hypothesis that the illusion is generated, at least in part, by so-called S1 cells in area V1. Most of these reasons involve compositional abduction. Schiller and Carvey cite features of the illusion that they claim cannot be explained by the RGC theory but can be explained by their S1 theory.

Before reviewing the abductive arguments, I should comment that Schiller and Carvey note that their paper does not report on experiments. They, instead, invite the reader to demonstrate the features of the illusion for herself: "Rather than reporting on experiments conducted in a small population of subjects, this paper consists of a series of demonstrations that allow the reader to serve as subject as well as judge. The demonstrations are compelling and we are confident they will convince the readers of the validity of the claims made" (Schiller & Carvey, 2005, p. 1376). Schiller and Carvey, along with the editors of the journal, believe that experiments are unnecessary. This work is, therefore, strictly speaking, not a matter of the interpretation of experimental results, but it is clear how the interpretation of the demonstrations would carry over to the interpretation of *bona fide* experimental results.

The lack of experimental results renders the Schiller and Carvey argumentation less than a perfect study for my purposes. Nevertheless, Schiller and Carvey (2005) does have the attractive feature that Schiller and Carvey are generally explicit in their appeals to explanatory considerations in support or criticism of a theory. For example, there is a section of the paper entitled "Why the retinal ganglion cell theory is untenable" and it begins with the contention that: "the retinal ganglion cell theory cannot appropriately explain the Hermann grid illusion" (Schiller & Carvey, 2005, p. 1377). Although such explicit commitments are not always forthcoming, it is convenient when there is a consilience of scientific and philosophical accounts.

Consider, now, three features of Hermann grid stimuli that Schiller and Carvey think challenge the RGC hypothesis but support a rival hypothesis. Here, I review only the first three of six features that do this work for Schiller and Carvey. The first three are the simplest and should suffice to make the point that Schiller and Carvey use abductive confirmation and disconfirmation. The interested reader can consult Schiller and Carvey (2005) for the remaining examples.

The illusion is perceived over a large range of sizes.[12] As Schiller and Carvey note, the reader might verify this feature of the illusion by viewing a familiar version of the illusion from different distances.[13] An experiment along these lines would move the viewer or the grid, and then measure the perceiver's perceptual response relying on the perceiver's verbal report. This would be an intralevel experiment. The RGC theory cannot explain the results of this experiment, since RGCs have fixed receptive field sizes. By contrast, Schiller and Carvey's rival hypothesis is that the illusion is generated by S1, simple cells, in area V1. These cells have much more variable receptive field sizes, which partly explains why the illusion is perceived over a range of sizes.

Note that the diversity of receptive field sizes is, for Schiller and Carvey, only part of the S1 theory's explanation of the illusion. In the first pass through the S1 theory's explanation, they write, "The fact that receptive field coverage is not limited to only one size can explain why, for a given retinal eccentricity, the illusion persists when the size of the grid is varied" (Schiller & Carvey, 2005, p. 1389). In a later commentary they write,

> [I]t has been shown that the illusion is effective over a range of sizes which is problematic for the retinal ganglion cell theory. The S1 theory is less affected by this observation for the following reasons: (I) the receptive fields

[12] The italicized headings here are borrowed from Schiller and Carvey (2005).

[13] Schiller and Carvey (2005, p. 1377).

> of cortical cells are larger and more variable in size . . . (ii) their responses are largely controlled by the orientation of edges . . . and (iii) responses in many cells summate as the length of the line segment is extended. (Schiller & Carvey, 2005, p. 1391)

These refinements do not change the central point that Shiller and Carvey think there is reason to believe in the S1 theory, namely, that it explains the variability in the size of the grids that stimulate the Hermann grid illusion.

The illusion is reduced when the grid is rotated by 45°. The reader might verify this observation by simply viewing the standard grid and then rotating it by 45°, but Schiller and Carvey refer to various experiments in the literature that support this view. They especially support a quantitative study of this phenomenon in De Lafuente and Ruiz (2004). The problem for the RGC theory is that the receptive fields of RGCs are radially symmetric, so the illusion should be unchanged by rotation of the grid. In contrast to RGCs, S1 cells are orientation sensitive. They are, one might say, oriented line detectors. Further, there appear to be more S1 cells dedicated to representing horizontal and vertical lines than there are to presenting diagonal/oblique lines. Thus, the S1 hypothesis can explain the weakening of the illusion in terms of fewer S1 cells responding to the grid.

The illusion can be reduced or eliminated by manipulations that do not alter the antagonistic center/surround activation of retinal ganglion cells. Here, Schiller and Carvey refer to a wide range of phenomena. For one thing, the illusion is reduced if the black squares forming the background are replaced by black parallelograms with four equal-length sides. As with the last problem, the problem for the RGC theory here is that the receptive fields of RGCs are radially symmetric, so the illusion should be unchanged by the angles of the white bars at the intersections. By contrast, the S1 theory can explain the weakening in terms of the fewer S1 cells representing the diagonals.[14]

6.5 Confirmation by Singular Interlevel Abduction

The cornerstone of my methodology is to carefully examine scientific reasoning regarding compositional hypotheses as presented in the primary experimental literature. Fidelity to actual science is the core desideratum. There are, however, other desiderata one might have. Many philosophers

[14] Surprisingly, Schiller and Carvey refer to other features of other grid variations that are not explained by the RGC theory, but for which they do not provide an S1 explanation.

of science seek a universal account. In the context of the confirmation of compositional hypotheses, this might be an account that applies to all instances of scientific reasoning in support of compositional hypotheses. In Chapter 1, I rejected the goal of such a universal account. In my reading of the scientific literature, there is such a diversity of things that are treated as explanations and used for abductive inference that, at this stage of philosophical investigation, it is unlikely that there will be a universal account. Here I will illustrate the scientific confirmation of a compositional hypothesis that does not use singular compositional abduction. The example relies on singular interlevel abduction.

Schepelmann et al. (1967) presented cats with Hermann grid stimuli and recorded responses from single cells. Given the RGC hypothesis, they anticipated finding cells that respond more strongly to the onset of lights on either horizontal or vertical bars, but more weakly to the onset of lights at the intersections of grids. Moreover, the findings were as expected. Here is Spillmann's commentary on the experiment:

> (Schepelmann et al., 1967) tested the receptive-field explanation of the Hermann grid illusion by recording impulses from individual fibers in the optic nerve and from single cells in the lateral geniculate nucleus and visual cortex of the cat. They found a neuronal response pattern consistent with the illusion for cells at all three levels, including simple cells. . . . The firing rate is about the same for the vertical and horizontal bars, but is much reduced when both bars are presented together, intersecting each other. This reduction of the firing rate might be considered a neurophysiological correlate of the perceived darkening; it is consistent with Baumgartner's receptive field explanation of the Hermann grid illusion. (Spillmann, 1994, p. 694)[15]

Notice that, in this test of the RGC theory, Schepelmann, Aschayeri, and Baumgartner intervene on cats, then measure responses from cells in the retina, lateral geniculate nucleus, and area 17. This is a top–down experiment, unlike the many intralevel experiments throughout Chapters 5 and 6. Such experiments will not be understood as are the previous intralevel experiments interpreted as compositional abduction.

Consider, once again, the first experiment from Spillmann (1971). (The second would work just as well.) In the first experiment, Spillmann found that when the bars are shortened, the illusion is weakened. This is a psychophysical result. It concerns the connection between a physical stimulus to an organism and the organism's perception. This result provokes an intralevel explanation-seeking why-question: "Why is the illusion

[15] Cf., Spillmann (2014, p. 1156).

weakened when the bars are shortened?" Spillmann's answer: The illusion is implemented by RGCs with ON-Center, OFF-surround receptive fields, so that as the cross bars decrease in size, they decrease their influence on the OFF-surround. Compare this with experiment in Schepelmann et al. (1967). In this experiment, Schepelmann et al. presented visual stimuli to cats, and then measured cellular responses. The result was not a standard psychophysical result; the experiment was not a standard psychophysical experiment. It did not manipulate a cat and then measure the cat's response; it manipulated a cat and then measured cellular responses. In other words, the Schepelmann et al. (1967) experiments were top–down experiments. The results of the Schepelmann et al. (1967) experiments, thus, provoke an interlevel explanation-seeking why-question: "Why do RGC cells respond more strongly to horizontal or vertical bars than to intersections?" Spillmann's answer: The illusion is implemented by RGCs with ON-Center, OFF-surround receptive fields, so that stimuli that produce the illusion also stimulate RGCs.

Both of the answers to the why-questions – both explanations – invoke the hypothesis that the illusion is implemented by RGCs with ON-Center, OFF-surround receptive fields. Both of these explanations are explanations of experimental results. Yet, in the former case, the explanandum is an individual engaged in an activity instance, whereas, in the latter case, the explanandum is of an interlevel connection. Thus, insofar as one types abductive inferences by reference to their embedded explanations, we have reason to draw a distinction between compositional abductive inferences and interlevel abductive inferences.

6.6 Summary

This case study illustrated many of the same points as made in the case study on resting and action potentials. In the interpretation of the results of controlled experiments in the primary experimental literature, scientists often use singular compositional abduction as a basis for confirmation and disconfirmation. One feature not found in the Hermann grid illusion study is the explanation of rates. Another, more significant, innovation from this case study is "singular interlevel abductive inference." These are abductive inferences based on an interlevel explanation. In the top–down experiment reviewed here, there was an interlevel explanation of an experimental result in which the presentation of a Hermann grid stimulus resulted in the increased firing of a retinal ganglion cell. I shall return to singular interlevel abductive inferences in Chapter 10.

CHAPTER 7

Peirce and the Neo-Peirceans

Although many see precursors to the idea of abduction in Aristotle's conception of *apagoge* or William Whewell's "consilience of inductions," the work of C. S. Peirce has had far and away the greater influence on philosophical thinking about abduction.[1] Indeed, there is a "Neo-Peircean" tradition that embraces many of Peirce's ideas with little question. Further, this tradition typically sets itself apart from the IBE tradition beginning with Harman (1965) and continuing most notably through Lipton (2003). This chapter focuses on the Peircean tradition, whereas the Chapters 8 and 9 focus on the IBE tradition.

I should emphasize at the outset that my goal is not to offer a critique of Peircean or Neo-Peircean views of abduction. Instead, my goal is to locate my view in philosophical space. I refer to Peirce and the neo-Peirceans to articulate my views. A primary rationale for this is that the Peircean and Neo-Peircean projects are largely orthogonal to mine. It would be a mistake to criticize Peirce and the Neo-Peirceans for not doing something they are not trying to do. Whereas my goal is characterizing how many scientists have reasoned about experimental results in twentieth-century scientific journal articles, Peirce's was not. While Peirce may have correctly articulated how he and his fellow scientists reasoned in the late nineteenth and early twentieth century, he obviously did not have in mind mid-twentieth-century scientific journals and their regimented formats for presenting experimental results. Nor did he have in mind compositional abduction. The Neo-Peircean attention to science is generally much more limited and indirect, typically by way of what other philosophers of science have had to say about science. Thus, Peirce and the Neo-Peirceans have not tried to focus their efforts on science as described in Chapter 1. Nor have they provided extended case studies of episodes in the history of science that they take to illuminate, or be illuminated by, their preferred

[1] Anderson (1986), Gabbay and Woods (2005), Bellucci and Pietarinen (2022).

conceptions.[2] Nor have they been concerned to develop a theory of explanation.

7.1 Peirce

If I were to give a succinct, rough comparison of my view with Peirce's, I would say that, whereas I accept the idea of abduction as an inference to an explanatory hypothesis, I find many of his other related ideas about abduction unhelpful for my project. A principal source of our divergences is what appears to be a bedrock Peircean assumption: scientists use abduction to introduce, but not confirm, new hypotheses.

Abduction, Deduction, and Induction. For present purposes, I will only discuss what is sometimes described as Peirce's "late" or "mature" theory of abduction. His 1901 work, "On the Logic of Drawing History from Ancient Documents," is sometimes identified as the first full treatment of this late theory.[3] Here, Peirce distinguishes between abduction, deduction, and induction both as patterns of inference and as stages in scientific inquiry. Let me begin with the description of abduction.

> Accepting the conclusion that an explanation is needed when facts contrary to what we should expect emerge, it follows that the explanation must be such a proposition as would lead to the prediction of the observed facts, either as necessary consequences or at least as very probable under the circumstances. A hypothesis, then, has to be adopted, which is likely in itself, and renders the facts likely. This step of adopting a hypothesis as being suggested by the facts, is what I call *abduction*. (Peirce, 1992, pp. 94–95, italics in original)

From this passage, I take the idea that abduction is not just any adopting of some hypothesis. It is adopting a hypothesis *because it is explanatory*. Adopting the hypothesis that sodium carries the initial inward current of the action potential, because it is a simple hypothesis would not be an abductive inference. In this passage, Peirce includes the idea that the explanandum should be "contrary to what we should expect." I do not include this in my understanding of abduction but will return to this issue.

I think Peirce and I share common ground in thinking of abduction as inferring a hypothesis because it is explanatory. Peirce next observes that it (often?) happens that for any explanation of some observation, there are

[2] The one exception may be Hanson (1972).

[3] Fann (2012, p. 32). See also Bellucci and Pietarinen (2022).

incompatible rival explanations so that a given hypothesis must be further tested and rivals eliminated:

> Aristotle had been a great reader of other philosophers, and it had struck him that there are various inconsistent ways of explaining the same facts. Ultimately, the circumstance that a hypothesis, although it may lead us to expect some facts to be as they are, may in the future lead us to erroneous expectations about other facts, – this circumstance, . . . was brought home to scientific men so forcibly, first in astronomy, and then in other sciences, that it became axiomatical that a hypothesis adopted by abduction could only be adopted on probation, and must be tested. (Peirce, 1992, p. 95)

Here again, Peirce and I share common ground. Scientists subject new hypotheses to subsequent tests. They do not merely pick one explanatory hypothesis and call an end to investigation.

As an account of this subsequent testing, Peirce offers hypothetico-deductivism:[4] "When this is duly recognized, the first thing that will be done, as soon as a hypothesis has been adopted, will be to trace out its necessary and probable experiential consequences. This step is *deduction* (Peirce, 1992, p. 95). Then, somewhat later, Peirce adds, "Having, then, by means of deduction, drawn from a hypothesis predictions as to what the results of experiment will be, we proceed to test the hypothesis by making the experiments and comparing those predictions with the actual results of experiment" (Peirce, 1992, pp. 96–97). Finally, "When, however, we find that prediction after prediction, notwithstanding a preference for putting the most unlikely ones to the test, is verified by experiment . . . we begin to accord to the hypothesis a standing among scientific results. This sort of inference it is, from experiments testing predictions based on a hypothesis, that is alone properly entitled to be called *induction*" (Peirce, 1992, p. 97). This gives us an abduction-deduction-induction conception of scientific method.

Where I break most decidedly with Peirce is in his contention that abduction and this hypothetico-deductivism are incompatible steps in a scientific method:

> The induction adds nothing. At the very most it corrects the value of a ratio or slightly modifies a hypothesis in a way which had already been contemplated as possible.

Abduction, on the other hand, is merely preparatory. It is the first step of scientific reasoning, as induction is the concluding step. Nothing has so

[4] McKaughan (2008, p. 451) concurs. See also Mohammadian (2021).

much contributed to present chaotic or erroneous ideas of the logic of science as failure to distinguish the essentially different characters of different elements of scientific reasoning; and one of the worst of these confusions, as well as one of the commonest, consists in regarding abduction and induction taken together (often mixed also with deduction), as a simple argument. (Peirce, 1992, p. 106)On my view, historians and philosophers of science do indeed need to distinguish abduction from the process of drawing out the consequences of a hypothesis and from the process of determining whether those consequences obtain. Abduction, deduction, and induction do involve distinct reasoning processes. Historians and philosophers of science should distinguish abductive, deductive, and inductive inferences. On my account, however, what Peirce describes scientists as doing in the deductive and inductive steps is, sometimes, scientists laying the groundwork for an abductive inference. Deduction and induction together are sometimes used for an abductive inference. I set out this view in Chapter 4, but I will revisit it here giving it a more Peircean gloss.

Bellucci and Pietarinen (2022) proposes a helpful example of how Peirce's three-stage process of abduction-deduction-induction works. However, I will use this example to illustrate how to reconcile my abductive approach with Peirce's comments on the deductive and inductive steps. Bellucci and Pietarinen's example begins with Peirce's early account of landing in a Turkish seaport:

> I once landed at a seaport in a Turkish province; and, as I was walking up to the house which I was to visit, I met a man upon horseback, surrounded by four horsemen holding a canopy over his head. As the governor of the province was the only personage I could think of who would be so greatly honored, I inferred that this was he. This was an hypothesis. (Peirce, 1878, p. 472)

Bellucci and Pietarinen propose that this example can be put in the "classic" 1903 schema:

The surprising fact, C, is observed.
But if A were true, C would be a matter of course.
Hence, there is reason to suspect that A is true (Peirce, 1992, p. 231),

as follows

The surprising fact that a man riding a horse surrounded by four horsemen holding a canopy over his head is observed.
But if this man were the governor of the province, the fact that he rides horses surrounded by four horsemen holding a canopy over his head would be a matter of course.

Hence, there is reason to suspect that this man is the governor of the province. (Bellucci & Pietarinen, 2022, pp. 5–6)

Bellucci and Pietarinen interpret the conditional as stating that there is a deductive relation between the man being the governor of the province and his riding a horse surrounded by four horsemen holding a canopy over his head. Further, the deductive relation is intended to offer an explanation: the man's being the governor of the province would explain the fact that he rides horses surrounded by four horsemen holding a canopy over his head. This much is the abductive step, the "first step in inquiry."

The second step is to trace further deductive consequences or predictions of the hypothesis that the man is the governor. Bellucci and Pietarinen offer the following:

This man is the governor of the province.
Governors of the province can speak French and wear long shirts under the coat.
Hence, this man can speak French and wears a long shirt under the coat. (Bellucci & Pietarinen, 2022, p. 8)

The third and final inductive step is to check whether the man can speak French and wears a long shirt under the coat. This might be done by addressing the man in French and somehow having him raise his coat. "[I]f the predictions are fulfilled, if he can replies [sic] in French and a long shirt becomes visible under the coat, the experiment has had [a] positive result, and the hypothesis that the man is the governor of the province is confirmed (at least, provisionally)" (Bellucci & Pietarinen, 2022, p. 9). This is simple hypothetico-deductive confirmation.

In Chapter 4, I proposed that scientific hypothetical reasoning that philosophers commonly interpret as hypothetico-deductive confirmation can, at least at times, also be interpreted as abductive reasoning. It is easy to see how to do this with Bellucci and Pietarinen's example. Rearrange matters to fit Peirce's "classic" abductive schema:

The surprising fact that this man can speak French and wears long shirts under the coat is observed.
If this man were the governor of the province, the fact that he can speak French and wears long shirts under the coat would be a matter of course.
Hence, there is reason to suspect that this man is the governor of the province.

As Bellucci and Pietarinen note, for Peirce, the deductive/explanatory relation between the man's being the governor, on the one hand, and his speaking French and wearing a long shirt under his coat, on the other, is the same deductive/explanatory relation as between the man's being the governor, on the one hand, and his riding a horse surrounded by four horsemen holding a canopy over his head, on the other.[5] This enables us to use the Peircean perspective to see hypothetico-deductive reasoning as a means of serving abductive inference.

One may, of course, observe that this is Bellucci and Pietarinen's example, so they can stipulate what happens in it. They can stipulate that the "scientists" in their example really do use HD confirmation and never reason abductively. That is fine by me, since I am not ultimately concerned with imaginary cases meant to serve expository purposes. I am instead interested in how scientists have reasoned in specific historical episodes. I offer a different theory of some of the actual scientific practices. So, let me review some real historical, scientific examples using the Peircean framing of the issues.

Consider representing Baumgartner's 1960 reasoning regarding the Hermann grid illusion and retinal ganglion cells according to Peirce's 1903 schema:

> The surprising fact that scintillating smudges are found in the Hermann grid is observed.
> But if RGCs implement the illusory perception, the fact that scintillating smudges are found in the Hermann grid would be a matter of course.
> Hence, there is reason to suspect that that RGCs implement the illusory perception.

Now consider Spillman's 1971 reasoning. Spillman envisioned that Baumgartner's hypothesis predicts that reducing the bar sizes would weaken the intensity of the scintillating smudges, so he did engage in what Peirce would call deductive-inductive reasoning. Yet, this reasoning serves to set up a further abductive inference that can be fit within Peirce's 1903 schema:

> The surprising fact that scintillating smudges in the Hermann grid are weaker when the bars are shortened is observed.

[5] Bellucci and Pietarinen (2022, p. 8).

But if RGCs implement the illusory perception, the fact that scintillating smudges in the Hermann grid are weaker when the bars are shortened would be a matter of course.

Hence, there is reason to suspect that RGCs implement the illusory perception.

This story respects much of what Peirce says about abduction-deduction-induction and scientific method. Spillmann did not simply accept Baumgartner's RGC hypothesis. Instead, he reasoned out one of its consequences, then performed an experiment to check the prediction and the experiment verified the prediction. Yet, I differ from Peirce in admitting that both Baumgartner's and Spillmann's inferences illustrate the idea of inferring some hypothesis because of what that hypothesis explains. This is why I say that scientists sometimes use abduction not merely to introduce some hypothesis, such as the RGC hypothesis. Instead, they may also confirm some hypothesis using abduction. Hypothetical reasoning does not conflict with abductive reasoning. Instead, hypothetical reasoning can set the stage for abductive reasoning.

Surprising facts. Peirce might object to the foregoing interpretation of hypothetico-deductivism as abductive on the grounds that, in the hypothetico-deductive phase, there is no surprising fact to be explained. Peirce might say that, in Bellucci and Pietarinen's example, it is not surprising that the man can speak French and wears long shirts under the coat, since this was a consequence deduced from the hypothesis that he was a governor. In the Spillman example, it is not surprising that scintillating smudges in the Hermann grid are weaker when the bars are shortened is observed, since this is a consequence of the RGC hypothesis. The time is now ripe to return to the issue of "surprising facts."

I noted above that, in "On the Logic," Peirce includes the idea that the explanandum should be "contrary to what we should expect." The more famous formulation of the idea is from the 1903 schema wherein the explanandum is "surprising." Adolfas Mackonis argues that "Peirce stated that abduction begins from a surprise... Thus, neither abduction nor IBE can be conducted if there is no surprising fact to be explained" (Mackonis, 2013, p. 977). Yet, Mackonis must have a suppressed premise linking what Peirce claimed about abduction and what is true of abduction. The obvious candidate is "If Peirce stated that abduction begins from a surprise, then neither abduction nor IBE can be conducted if there is no surprising fact to be explained." But why believe this premise? Is the assumption here that abduction just is what Peirce says it is?

Jaime Nubiola has gone as far as proposing that abduction is the logic of surprise.[6] Of the 1903 schema, Nubiola writes, "This is the logical structure of all abductions. The key for understanding it properly is to realize that the trigger of abduction is the surprising character of the fact referred to in the first premise" (Nubiola, 2005, p. 126). For my part, I am willing to allow that Peirce's 1903 schema captures the logical structure of all *Peircean* abductions and that the key for understanding *Peirce's* conception of abduction properly is to realize that the trigger of abduction is the surprising character of the fact C. In this book, however, I am proposing a conception of abduction that does not presuppose this feature.[7] Nubiola might not count my "thin" conception of abduction as a Peircean conception. He might – like Mackonis – not count my "thin" conception of abduction as a conception of abduction at all. Nevertheless, I persist in this conception because my non-Peircean conception of abduction better describes the cases I examine.

Consider the point more concretely. Scientists often design experiments with preconceptions regarding the future results of those experiments. They may do this after what Peirce described as deductive reasoning. Scientists are guided in these preconceptions by their theories. In many of these cases, they are not surprised by the results. Thus, after learning of the results of Blodgett's experiment with the six-unit T-maze, Tolman was probably not surprised to find similar results with the fourteen-unit T-maze. Given the latent learning in the six-unit T-maze, Tolman was not surprised to find latent learning in the fourteen-unit T-maze. Similarly, given the currents discovered in their first two experiments on the sodium hypothesis in Hodgkin and Huxley (1952a), Hodgkin and Huxley had some definite ideas about what would happen in their third experiment wherein the axon was voltage-clamped in a medium containing 10% and 30% of the normal sodium of content of seawater.

Tolman believed that the rat navigates with a cognitive map, since that would compositionally explain the rat's performance in the maze. This inference is (thin) compositional abduction. Hodgkin and Huxley believed that sodium carries the initial inward current of the action potential, since that would compositionally explain the currents in the reduced sodium media. This inference is (thin) compositional abduction. The theory of compositional abduction captures this feature of Tolman's and Hodgkin and Huxley's scientific writing.

[6] Nubiola (2005). [7] Gabbay and Woods (2005, p. 85) drop the surprise condition as well.

Abduction is for Generating Hypotheses; IBE is for Confirming Hypotheses. There is a tradition in Peircean and Neo-Peircean scholarship to propose, to a first approximation, that Peirce's concept of abduction assumes that abduction is for introducing/generating hypotheses, whereas Harman's or Lipton's IBE is for hypothesis confirmation.[8] This claim is often juxtaposed with the idea that abduction – with no mention of Peirce – is for introducing/generating hypotheses, whereas IBE – with no mention of Harman or Lipton – is for hypothesis confirmation.[9] So, for example, Daniel Campos writes,

> I argue against the tendency in the philosophy of science literature to link abduction to the inference to the best explanation (IBE), and in particular, to claim that Peircean abduction is a conceptual predecessor to IBE. This is not to discount either abduction or IBE. Rather the purpose of this paper is to clarify the relation between Peircean abduction and IBE in accounting for ampliative inference in science. This paper aims at a proper classification – not justification – of types of scientific reasoning. In particular, I claim that Peircean abduction is an in-depth account of the process of generating explanatory hypotheses, while IBE, at least in Peter Lipton's thorough treatment, is a more encompassing account of the processes both of generating and of evaluating scientific hypotheses. There is then a twofold problem with the claim that abduction is IBE. (Campos, 2011, p. 419)

Campos is clearly shifting back and forth between writing about abduction and about Peircean abduction. Of course, insofar as this is a mere stylistic shift, there is no philosophical issue at stake. And plenty of scholars are careful not to make this stylistic shift.[10] There is, however, a philosophical issue if one assumes that Peirce's concept of abduction somehow constrains any theory of abduction. There is a philosophical issue if one contends that abduction just is what Peirce says it is. This is an assumption that requires further argumentation than it has been given in the literature.[11] I am happy to allow that my conception of abduction differs from Peirce's – maybe it is a "non-Peircean" concept of abduction, but it is quite another matter to hold that what I call abduction is not abduction at all.

8 McKaughan (2008) proposes a third alternative: abduction introduces hypotheses worthy of pursuit. See also Nyrup (2015).

9 There are other ways of reading Peirce as well. McKaughan (2008) proposes a third option, that abduction introduces hypotheses as worthy of pursuit. Magnani (2001, p. 25) suggests, without textual references, that Peirce sometimes treated abduction as for both hypothesis generation and hypothesis confirmation.

10 McKaughan (2008), Cabrera (2022).

11 See, for example, McAuliffe (2015) and Mohammadian (2021).

More concretely, one might think that since abduction is for hypothesis generation/ introduction, whereas IBE is for hypothesis confirmation, the very idea of abductive inference for hypothesis confirmation is a conceptual confusion. My reply is that I am not using the Peircean concept of abduction. I am developing another one – taking one element from the Peircean view – that I believe supports a more detailed and accurate description of scientific reasoning in the primary experimental literature than does the Peircean conception. I am not confused at all.

One might, however, object that my terminology is confusing to others. Yet clearly the terminology is only potentially confusing to those who are resolutely locked into the picture of abduction for hypothesis generation and IBE is for hypothesis confirmation. Philosophers of science outside the Neo-Peircean tradition should not be confused. Further, suppose one holds the view that abduction is for hypothesis generation and IBE is for hypothesis confirmation. Nevertheless, if one can at least conceive that some philosophers hold a different view and argue for that different view, then there should be no confusion.

It should be emphasized that almost all the discussion of this topic in Peirce scholarship focuses on comparing Peircean ideas with one or another conception of IBE. No serious consideration is given to the extent to which either conception correctly describes scientific practice. There is no serious attempt to develop a science-first interpretation of abduction or IBE. This is one reason why the project of understanding Peirce's view need not compete with my project of describing how scientists confirm compositional hypotheses.

7.2 Neo-Peirceans

I introduce "Neo-Peircean" as a term for those philosophers who begin with Peirce's idea of abduction as a means of hypothesis generation/ introduction. A Neo-Peircean might then show how Peirce's concept applies in one or another area of intellectual activity. A Neo-Peircean might introduce additional distinctions. Or a Neo-Peircean may try to build a formal theory of Peircean abduction. Insofar as these approaches address scientific practice – which they often do not – they embody a "philosophy first" approach, taking pre-existing philosophical ideas and applying them to scientific cases. This approach is diametrically opposed to the science-first approach I have taken in Parts 1 and 2 of this book. Little, if any, of the "philosophy first" work shows how Neo-Peircean ideas apply in detailed case studies. Moreover, none of this Neo-Peircean work develops a theory of explanation. I would classify most papers in the monumental *Handbook of Abductive Cognition* as "Neo-Peircean."[12] Rather than attempting to discuss the large Neo-Peircean literature, I will instead briefly touch on two prominent Neo-Peircean contributions, namely, Lorenzo Magnani's *Abduction, Reason, and Science* (Magnani, 2001), and Gabbay and Woods's *The Reach of Abduction* (Gabbay & Woods, 2005).

7.2.1 *Magnani*

Magnani (2001) is a paradigm of Neo-Peircean work. Following Peirce, Magnani assumes that abduction is for hypothesis generation. Chapter 1 of his book sets the stage for this by reviewing Platonic, Kantian, and Polanyian philosophical theories of hypothesis generation. Magnani's Chapter 2 reviews, among other things, a version of the familiar idea of abduction for hypothesis generation, IBE for hypothesis confirmation, with Magnani preferring abduction for hypothesis generation. It also embraces a form of hypothetico-deductivism.

Chapter 2 introduces what is widely considered one of Magnani's most important distinctions, between selective and creative abduction. Both abduction and IBE can be classified as either selective or creative. In brief, a selective inference selects some hypothesis from an existing stock of possible explanations. So, in Magnani's preferred context of medical diagnosis, a physician might use either abduction or IBE to select some hypothesis H_i from a pre-existing set of possible hypotheses, H_1, H_2, . . ., H_n, because H_i explains the symptoms some patient displays. By contrast, in a creative abduction, a physician/scientist might use either abduction or IBE to advance a new hypothesis regarding a new disease that explains some patient's symptoms.

Let me explain why Magnani's selective/creative distinction is of limited use for my project. Consider, first, the Hodgkin–Huxley case study. By 1947, Hodgkin and Huxley had arrived at the hypothesis that sodium ions carried the initial inward current of the action potential. Suppose for the sake of argument that Hodgkin and Katz's replication of the Overton experiment was creative. Once that was performed there were additional experiments, the experiments with reduced-sodium and enriched-sodium media, that were not creative. They were not creating a new hypothesis. It had already been created. So, creative abduction for hypothesis

[12] Magnani (2023).

generation does not illuminate why Hodgkin and Huxley did the reduced-sodium and enriched-sodium experiments. Further, neither selective abduction nor selective IBE applies, as the sodium hypothesis had already been selected.

Next, consider again Baumgartner's 1960 reasoning and Spillmann's 1971 reasoning. Suppose, for the sake of argument, Baumgartner offered a creative abductive inference generating the hypothesis that RGCs implement the grid illusion. Spillmann subsequently tested the hypothesis by reasoning hypothetically that the grid with shorter bars would produce a weaker illusion. This prediction was confirmed. I have proposed this hypothetical reasoning to be understood as compositional abduction. Yet, Spillmann's reasoning was neither creative nor selective. The RGC hypothesis had already been introduced by Baumgartner and Spillmann was not trying to select the RGC hypothesis from rivals.

Finally, consider Tolman's reasoning about the six-unit and fourteen-unit T-mazes. Suppose, for the sake of argument, that the cognitive map hypothesis was introduced in response to the six-unit maze results. When Tolman and his colleague did the replication with the fourteen-unit maze, they did not introduce a creative new hypothesis to explain the results. They were further testing the cognitive map hypothesis. Moreover, they were not selecting the cognitive map hypothesis from among rivals. They had already fixed upon the cognitive map hypothesis.

Next, consider a second of Magnani's major distinctions, the distinction between theoretical and manipulative abduction. He writes, "theoretical abduction is the process of *inferring* certain facts and/or laws and hypotheses that render some sentences plausible, that *explain* or *discover* some (eventually new) phenomenon or observation; it is the process of reasoning in which explanatory hypotheses are formed and evaluated" (Magnani, 2001, pp. 30–31, italics in original). This looks to be a standard understanding of abduction as the process of inferring some hypothesis because it is explanatory, albeit allowing that this inference may be formative or evaluative. Although Magnani introduces the term "manipulative abduction" early in his book and comments on it at several points, he only begins articulating what he means by it in Chapter 3. Magnani's description of manipulative abduction is less clear to me, so I quote him at length:

> Manipulative abduction ... happens when we are thinking through doing and not only, in a pragmatic sense, about doing. So the idea of manipulative abduction goes beyond the well-known role of experiments as capable of forming new scientific laws by means of the results (the nature's answers to the investigator's question) they present, or of merely playing a predictive

> role (in confirmation and in falsification). Manipulative abduction refers to an extra-theoretical behavior that aims at creating communicable accounts of new experiences to integrate them into previously existing systems of experimental and linguistic (theoretical) practices. (Magnani, 2001, p. 66)

I take from this that a manipulative abduction is an inference that happens more or less spontaneously, as when scientists physically manipulate objects in the laboratory.

Here again, this distinction is largely orthogonal to my project. Given a forced choice between Magnani's categories of theoretical and manipulative abduction, I would have to say that all my examples are instances of theoretical abduction. The scientists who I have discussed may well have engaged in manipulative abduction in the laboratory, but my project is not intended as an account of the psychology of laboratory life. It is more like a sociological account of how scientists interpret certain experimental results to support hypotheses in the primary experimental literature. What makes it sociological, rather than psychological, is that it is shaped by social conventions about how scientists present experimental results and interpret them for their fellow scientists within the confines of relatively regimented scientific journal articles. This reasoning is somewhat removed from a completely faithful recounting of scientific thinking in the laboratory. Part of the burden of Chapter 1 was to emphasize this feature of my account.

The foregoing is not, of course, a critique of Magnani's project. He has not directed his efforts as I have. He takes on a lot of the Peircean framework and adds some further refinements, but he is not engaged in a science-first project in the philosophy of science. He is not, in the first instance, aiming to use scientific examples to build a theory of compositional abductive inference. Instead, Magnani tells us that "the aim of this book is to emphasize the significance of abduction in order to illustrate the problem-solving process and to propose a unified epistemological model of scientific discovery ..., diagnostic reasoning ..., and other kinds of creative reasoning (Magnani, 2001, p. 15). Magnani's stated aim differs from mine at multiple levels. It should be no surprise, therefore, that the conceptual tools Magnani develops for his project are of little use to mine.

7.2.2 *Gabbay and Woods*

Gabbay and Woods (2005) is a monumental contribution to the logic of abduction. Moreover, it begins with a promising nod to what looks like a science-first philosophy of science and logic of abduction: "In recent decades, philosophy of science has evolved from its original stance to its present non-foundational or *neoclassical* [stance] . . . Accordingly, the flow of inquiry is bottom-up, from science, to the philosophy of science, to the logic of science" (Gabbay & Woods, 2005, p. 4). Yet, quite quickly, the science-first approach is abandoned. When describing the targets of their formalization, Gabbay and Woods write, "In a rough and ready way, these can only be what is known informally about how agents actually reason. In a well-entrenched analytic tradition, inputs are made accessible by the theorist's intuitions about the subject at hand, which stand as his *data* for a theory of it. Intuitions are the conceptual data for theory, the inputs to the theory's formal models" (Gabbay & Woods, 2005, p. 4). Yet, neither a logician nor a historian and philosopher of science needs to rely on what is informally known about how agents actually reason. The input to the theory need not be a theorist's "intuitions." Instead, one might make a close study of actual scientific reasoning, maybe using cognitive science or history and philosophy of science. Perhaps looking to cognitive science or history and philosophy of science only leaves one with an "informal" account, but it is also not merely looking to a logician's intuitions. So, right away, Gabbay and Woods undertake a project that is orthogonal to mine. It is not a science-first project.

A second feature of Gabbay and Woods's project is that they aspire to something more general than a logic of scientific reasoning. "As conceived of in the neoclassical tradition, the logic of science provides idealized models of the most general and central features of the conceptual content of the process of scientific reasoning. If we remove the qualification 'scientific', the logic which the new classical logic of science exemplifies resembles our own more general conception of logic" (Gabbay & Woods, 2005, p. 4). Gabbay and Woods aspire to something like a "general theory of reasoning." My project is near the opposite end of a spectrum. As I have been at pains to note at many points, singular compositional abduction is only one form of scientific reasoning. It is, indeed, only one form of abductive reasoning. Moreover, this book offers a granular reading of some of the scientific literature, focusing on details of the scientific interpretation of specific experimental results. Indeed, one might naturally view this granular work, rather than a logician's intuitions, as a reasonable starting point for developing a logic of abduction. In principle, the specific scientific cases might be a basis for some sort of generalization to the philosophy of science, then to the logic of science, and then to logic in general.

This drive for generality creates some tension for historians and philosophers of science trying to understand the reasoning by Baumgartner,

Hodgkin, Huxley, Spillmann, and Tolman. On the one hand, Gabbay and Woods propose that abduction as an inference to an explanatory hypothesis is a mere special case of abduction.[13] This special case is what they call explanatory abduction. For Gabbay and Woods, abductively inferred hypotheses are pragmatically justified for various reasons, only one of them being that the hypothesis is explanatory. This is a manifestation of a drive for generality. It suggests that the reasoning by Baumgartner, et al., will fall out as a very special case of abduction. On the other hand, Gabbay and Woods also propose a strong narrowing of the concept of abduction: "As we see things here, the fundamental fact about abduction is that abduction is *ignorance-preserving reasoning*" (Gabbay & Woods, 2005, p. 40). As a conceptual point, abduction cannot lead to knowledge. So, we may face a forced choice: either Hodgkin and Huxley did not know that the sodium carries the initial inward current or they were not using abduction in the Gabbay and Woods sense. Maybe the simplest resolution is to say that Hodgkin and Huxley were not using Gabbay–Woods abduction.

Some of the Gabbay–Woods discussion of explanatory abduction has an interesting feature for the present work. (Not the discussion of explanation, which is largely a review of the DN model and some familiar alternatives.) Gabbay and Woods introduce a schema for explanatory abduction along the following lines:

1. E
2. It is not the case that K $\hookrightarrow$ E.
3. It is not the case that H $\hookrightarrow$ E.
4. K(H) is consistent
5. K(H) is minimal
6. K(H) $\hookrightarrow$ E.
7. Therefore H (cf., Gabbay & Woods, 2005, pp. 48–49),

Where E stands for some evidence, K is a set of background beliefs, H is a putative explanatory hypothesis, and $\hookrightarrow$ is an explanatory consequence relation. What is interesting is that Gabbay and Woods treat entailment as one instance of the explanatory consequence relation, so that their take on explanatory abduction looks very much like a version of HD confirmation.[14] Of course, on my account, the mere logical features of this schema and of HD confirmation do not capture the worldly thinking that I propose engages practicing scientists.

[13] Gabbay and Woods (2005, p. 40). [14] See, for example, Glymour (1980b).

Yet again, as with Magnani's treatment of abduction, Gabbay and Woods do not develop a set of tools germane to my project. They do not seriously look to scientific practice. Nor do they develop a theory of explanation. While one might hope that the formal tools would offer some theoretical clarity, the foundational assumption that abduction is ignorance-preserving means that the account fails to capture one of the key features of the scientific use of explanation to confirm scientific hypotheses.

7.3 Summary

The gist of this chapter has been to argue that there is but a slender connection between my work and that of Peirce and the Neo-Peirceans. This connection is captured in the idea that abduction is an inference to some hypothesis because of what that hypothesis explains. Whereas I believe that compositional abduction is sometimes used for scientific confirmation, Peirce and the Neo-Peirceans believe abduction is for hypothesis introduction. This difference has further consequences for the features one takes abductive inference to have. Most notable are the questions of how historians and philosophers of science are to understand scientific hypothetical reasoning, how surprising the abductive explananda are, how novel the explananda are, and what confirmation, if any, is provided by an abductive inference.

As a final word, one might suggest that I not use Peirce's word "abduction." Using "abduction" is confusing. Yet, a new word choice has its challenges. Maybe call the view "explanationism." Even with the new term, the question would still arise regarding how explanationism relates to the Peircean concept of abduction. One might also object that this new terminology slights Peirce by not acknowledging his role in championing the idea of an inference to some hypothesis because of what that hypothesis explains. Further, one might ask how my explanationism relates to the Harmanian/Liptonian concept of IBE. Maybe there is still another term I could use, but the principal issue remains. How does my concept relate to Peirce's abduction and Harman's IBE? There is no avoiding the project of philosophical orientation. I stand by "abduction" as a sensible terminological choice.

CHAPTER 8

Harman and Inference to the Best Explanation

In Chapter 7, I drew attention to a putative forced conceptual choice. Some philosophers have proposed that abduction is for hypothesis generation, whereas IBE (inference to the best explanation) is for hypothesis confirmation. Given this forced choice, one might think that, since my view is that scientists sometimes use abduction for confirmation, I should say that scientists use IBE. That would be clearer.

This forced choice, however, runs entirely counter to my science-first approach to understanding scientific reasoning. My goal is not to select one of these two preexisting conceptions, Peircean abduction versus Harmanian IBE, to impose on scientific cases. That would be a philosophy-first approach. My goal is to use scientific cases as a basis upon which to build an account of scientific interpretation. In Chapter 7, I argued that certain Peircean and Neo-Peircean ideas, developed outside the context of close attention to the mid-twentieth-century primary experimental literature, do not accurately describe the reasoning of Baumgartner, Hodgkin, Huxley, and Tolman, among others. Similarly, in this chapter, I argue that Harman's ideas, developed with no particular interest in the primary scientific experimental literature, do not helpfully describe this reasoning either. Harman's view may be sufficiently general or open-ended as to include singular compositional abduction, but that very generality or open-endedness means that it lacks the detail one might want in a philosophical account of how scientists interpret controlled experiments.

I must emphasize that my goal is not to "refute" Harman's IBE model. To do that would presumably require an account of what IBE is, which is tangential to my concerns. So, for example, one might think that IBE requires that there be a set of hypotheses $H_1, H_2, \ldots, H_n$ from which one selects one, H_i, that has the best weighted balance of virtues, such as simplicity, plausibility, and scope. Then again, one might maintain that the case in which there is only one hypothesis is just a special case of IBE.[1] Far be it from me to tell advocates of IBE what IBE is.[2] One might also think that, since Harman proposed that IBE is warranted abduction, then, at the very least, by definition, IBE must be a warranted inference. Then again, one might think that, in typical cases where IBE might be involved, one cannot eliminate all the rival hypotheses, so that in typical cases, one does not have warrant, but one still has something describable as IBE.[3] Again, far be it from me to tell advocates of IBE what IBE is. A subproject of specifying what IBE is, is tangential to my concerns. My overarching concern is developing an account of how scientists interpret the results of controlled experiments. In this chapter, I only intend a bit of philosophical cartography, indicating how my view relates to some of the ideas in Harman (1965).

8.1 IBE as Warranted Abduction

Harman's "Inference to the Best Explanation" is among the most influential papers in the entire abduction literature, spawning many disparate lines of philosophical inquiry. The principal thesis of the paper is that warranted enumerative induction is a special case of IBE. A secondary thesis is that taking IBE as the basic form of nondeductive inference accounts for an interesting feature of using the word "know." In contrast to these two central theses, there are passing "stage-setting" comments that have taken on a life of their own in the philosophical literature. These include discussions of (a) the relationship between Peirce's concept of abduction and Harman's concept of IBE; (b) a role for explanatory virtues, such as simplicity; and (c) the idea that IBE is used in science. My concern in this section is to focus philosophical attention on Harman's two principal theses and the stage-setting comments, thereby enabling me to indicate how the picture of IBE Harman developed for his project need not be deployed in my project of developing an account of how scientists interpret experimental results in support of compositional hypotheses.

Harman (1965) is a tightly argued paper that surely repays reading. It begins with a statement of Harman's primary objective: "I wish to argue

[1] See McCain (2014, p. 68).
[2] For efforts along these lines, see for example Lipton (2003), McCain (2015), and Lange (2022).
[3] Bird (2022, p. 177) calls this "ordinary IBE", as opposed to what he calls "Holmesian inference" or "inference to the only explanation."

that enumerative induction should not be considered a warranted form of nondeductive inference in its own right. I claim that, in cases where it appears that a warranted inference is an instance of enumerative induction, the inference should be described as a special case of another sort of inference, which I shall call 'the inference to the best explanation'" (Harman, 1965, p. 88, cf. p. 91). The gist of Harman's argument is simple: An enumerative induction from "All observed A's are B's" to "All A's are B's" is warranted when the hypothesis that all A's are B's is the best explanation of why all the observed A's are B's.

Harman's secondary objective is to show that there is a feature of knowledge that is captured within the framework of IBE that is not captured in terms of enumerative induction. The feature is that our knowledge is often thought to rely on what Harman calls "lemmas" or what might be called "background knowledge." Take an example of knowledge by testimony. Suppose Jones believes that P because she read that P in a scientific journal article. To thereby know that P requires background knowledge that P does not include a falsifying typographical error. In this case, mere background *belief* that P does not include a falsifying typographical error would not suffice for knowing that P. According to Harman, IBE captures this feature. He proposes that the best explanation of the journal article's containing P is that P was written without a typographical error. With enumerative induction, however, there is no role for the lemma. Instead, on an enumerative induction, the inference can only be of the form:

the journal printed P* and P* was true,
the journal printed P** and P** was true, etc.,

so that one can infer that when the journal printed P that P is true. There is no role for background knowledge regarding a typographical error in the enumerative induction story.

Here, I do not mean to defend Harman's account. Instead, my primary point is to contrast Harman's project with my project of correctly describing a bit of scientific practice. A key element here is that Harman was interested in warranted inference. Further, I wish to emphasize how some of Harman's "stage-setting" comments have shaped the later philosophical literature. These are his passing comments about the relationship between abduction and warranted abduction (Section 8.2), the role of explanatory virtues (Section 8.3), the role of IBE in science (Section 8.4), and the connection between abduction and IBE (Section 8.5).

8.2 The Ramifications of Warrant

As noted above, Harman's goal was to defend the epistemological thesis that warranted enumerative induction is a special case of what he called "inference to the best explanation": "in cases where it appears that a warranted inference is an instance of enumerative induction, the inference should be described as a special case of another sort of inference, which I shall call "the inference to the best explanation" (Harman, 1965, p. 88). A key element here – one that is not always made explicit in subsequent discussions – is warrant. Indeed, Harman refers to warrant twice more in his brief introductory section.

In introducing the concept of IBE, Harman explicitly connects warrant with the elimination of rival explanatory hypotheses:

> In making this inference one infers, from the fact that a certain hypothesis would explain the evidence, to the truth of that hypothesis. In general, there will be several hypotheses which might explain the evidence, so one must be able to reject all such alternative hypotheses before one is warranted in making the inference. Thus one infers, from the premise that a given hypothesis would provide a "better" explanation for the evidence than would any other hypothesis, to the conclusion that the given hypothesis is true. (Harman, 1965, p. 89)[4]

Harman's point is that a rival explanatory hypothesis would undermine one's justification or warrant. A rival hypothesis would be a defeater. Thus, in order for one's preferred hypothesis to be warranted, the defeaters must be removed. It is this elimination of all the rivals that motivates Harman to describe warranted abductive inference as "inference to the best explanation." Again, a foundational motive in formulating abduction as IBE is the concern with warrant.

For many subsequent projects, philosophers have not needed to distinguish between abduction and warranted abduction. So, Alexander Bird once proposed to use "inference to the best explanation" as a synonym for "abductive inference," rather than as a synonym for "warranted abductive inference."[5] In proposing this usage, I do not think Bird intended to offer a substantive philosophical thesis. Nor do I think he intended to register some comment on Peirce's view of abduction. He simply set out

[4] Harman is, of course, not the first to note an important role for rival hypotheses. Peirce (1992) noted it, but so did Frankfurt (1958).

[5] Bird (2005, p. 5).

terminology that suited his local philosophical purpose, namely, to articulate a concept of "Holmesian inference."

Douven (2017a) provides an authoritative account of abduction, but one that does not mark a distinction between warranted and unwarranted abduction. Like Harman, Douven is concerned with warranted abduction. But, when broaching a schema for abduction, he labels the relevant section of his discussion "Explicating Abduction," rather than "Explicating Warranted Abduction." Moreover, he supposes that the core idea of abduction is that "explanatory considerations have confirmation-theoretic import, or that explanatory success is a (not necessarily unfailing) mark of truth" (Douven, 2017a, p. 11). The implication of warrant is implicit at best.

One philosophical interest in warranted abduction/IBE is as a type of argument for scientific realism.[6] Many philosophers of science wish to argue that the best explanation for the success of science is that scientific terms refer and scientific hypotheses are true. In this context, the philosopher of science is not merely interested in abductive inference, the philosopher of science is interested in warranted abductive inference. In the context of the scientific realism debate, there is little need to be concerned with the difference between abduction and warranted abduction.

As this is such a deeply embedded assumption in the IBE tradition, I will spend a lot of time trying to spell out how failure to distinguish between abduction and warranted abduction can lead to some dubious history of science. So, to begin, Lipton writes, "Of course, there is always more than one possible explanation for any phenomenon – the tracks might have instead been caused by a trained monkey on snowshoes, or by the elaborate etchings of an environmental artist – so we cannot infer something simply because it is a possible explanation. It must somehow be the best of competing explanations" (Lipton, 2003, p. 56). I assume that Lipton's claim regarding what one cannot infer is not about the limits of our psychological capacities, that certain inferences are beyond the capacities of our finite minds. Instead, I take him to mean that one cannot *justifiably* infer something simply because it is a possible explanation. What one can justifiable infer, however, does not tell us how scientists actually infer.

Schurz tells a similar story at greater length:

> [I]f your evidence consists in the trace of the imprints of sandals on an elsewhere empty beach, then your immediate conjecture is that somebody was recently walking here. How did you arrive at this conjecture? Classical

[6] For a small sample, see Boyd (1983), Day and Kincaid (1994), Psillos (2005), Mizrahi (2012), Doppelt (2014), and Bird (2020).

> physics allows for myriads of ways of imprinting footprints into the sand of the beach, which reach from cows wearing sandals on their feet to footprints which are drawn into the sand, blown by the wind, or caused by radioactive decay of foot-shaped portions of the sand, etc. The majority of these physically possible abductive conjectures will never be considered by us because they are extremely improbable. The major strategic algorithm which we apply in factual abduction cases of this sort is a probabilistic elimination technique which usually works in an unconscious manner: our mind quickly scans through our large memory store containing millions of memorized possible scenarios and only those which have minimal plausibility pop up in our consciousness. (Schurz, 2008, p. 207)[7]

Schurz asks how his hypothetical beachcomber came up with the conjecture that somebody was recently walking on the beach. There is, no doubt, some psychological story to be told. Moreover, classical physics does indeed allow for a myriad of alternatives. However, just because there is some psychological story to be told and physics allows for alternatives to someone walking on the beach does not provide any reason to think that the beachcomber gave any of the alternatives the slightest consideration, conscious or otherwise. What goes on in the beachcomber's mind is an empirical question. Schurz assumes that the psychological story includes subconsciously entertaining some of the millions of alternative hypotheses allowed by classical physics. But why think that? What are these millions of possible scenarios? And why think there are millions of them, rather than dozens, hundreds, or billions? Empirical evidence about the case is needed, but Schurz provides none. What is driving Schurz is a concern with warranted inference. This, in turn, drives him to a speculative psychological proposal.

Here is another perspective on the speculative psychology. Suppose one asks the beachcomber, "How do you know that the marks in the sand were not caused by radioactive decay of foot-shaped portions of the sand?" The beachcomber might say that radioactive decay of such a shape is highly improbable. Suppose that suffices to rule out the radioactive decay hypothesis. However, the beachcomber might also say, "What is radioactive decay?" Here, the beachcomber has no concept of radioactive decay, hence could not have ruled out the radioactive decay hypothesis, either consciously or unconsciously. Philosophers of science interested in specific instances of actual scientific reasoning should not assume *a priori* that

[7] Schurz's discussion is nice, since it compactly presents the principal ideas. Similar discussions, however, can be found in Okasha (2002, p. 23f) and McCain (2019, p. 55f).

actual cases will proceed along the lines of the first scenario rather than the second.

One might observe that it is Schurz's imaginary example, so he can imagine it as he pleases. And indeed, he can. Or one might observe that Schurz is merely trying to explain what IBE reasoning is like. And maybe the latter is the more charitable reading of his comment. My concern, however, is to draw attention to how historians and philosophers of science should *not* relate imaginary examples, such as Lipton's or Schurz's, to actual scientific examples. While Lipton's or Schurz's purposes may require that an agent somehow eliminate rival hypotheses, mine do not.

The need for an abduction/warranted abduction distinction is relevant to *bona fide* scientific cases. Sometimes, scientists respond to the proposal of a rival hypothesis by pointing to features of their experimental design that rule out the rival. The rat could not be following an odor trail, since the arms of the maze were cleaned between each trial. Sometimes the rival might be ruled out by a prior experimental result. However, sometimes scientists also respond to a rival hypothesis by saying, "That is a good point. I should run an experiment that controls for that." Philosophers of science who are sufficiently familiar with actual science will know of representative cases. Thus, I propose *not* to assume *a priori* in any given case that scientists are making warranted abductive inferences. So, I am not assuming *a priori* that they have somehow eliminated some rival hypotheses. I assume that, in any given case, which rival hypotheses, if any, scientists may have eliminated is an empirical question.

What do I say to those philosophers who say that abduction is synonymous with the examination of rival hypotheses?[8] These philosophers are counterparts to Peirce scholars who insist that abduction is simply synonymous with inference to a novel explanatory hypothesis. I would suggest that this usage is stipulative and that I am not using abduction in that sense. Moreover, I would note that there is a long tradition of using "abduction" to describe a defeasible, inductive inference to some hypothesis H on the ground that H explains E, without the presupposition that the inference somehow involves some perhaps tacit consideration of rivals to H. Recall from Chapter 7 that Peirce's "Turkish province" story involved no rival hypotheses.

In distinguishing between abduction and warranted abduction, I might cite a similar precedent. Norton discusses two examples of what he takes to be enumerative induction:

[8] I do not assume that Bird would take this line, but in point of logic he might.

Some men are mortal.
Therefore, all men are mortal.

Some men are Greeks.
Therefore, all men are Greeks.[9]

He considers both of these instances of enumerative induction, but argues that the first is justified, whereas the second is not. Norton does not build a requirement of warrant or justification into the very idea of enumerative induction. Just so, I propose to use the term "abduction" so as not to build in a requirement of warrant, hence not build in a requirement that rival hypotheses be considered and eliminated.

At this point, one might object, "By setting aside questions regarding the warrant of abductive inferences, are you not casting doubt on the warrant of science, or at least the warrant of the parts of science that you think scientists have confirmed abductively?" No. For it is one thing to withhold judgment on the warrant of individual inferences at particular points in time, but quite another to withhold judgment on the warrant of the entire body of scientific reasoning. To suspend one's judgment on the warrant of Hodgkin and Huxley's first 1952 inference to the sodium hypothesis when it was first advanced says nothing about the warrant the hypothesis eventually comes to have. To return to a formulation I gave earlier, it might be that such warrant as the sodium hypothesis enjoys is not the product exclusively of the interpretation of the first Hodgkin–Huxley experiment, but instead the product of a temporally extended process of examining and eliminating multiple rivals to the sodium hypothesis. Maybe a scientist eliminates one rival based on one experiment, then a few pages later eliminates another rival based on another experiment. Furthermore, one might well endorse the overall warrant of scientific reasoning in some global sense without having to endorse every single inference made by every scientist. Indeed, I would speculate that any philosopher of science who reads enough science will eventually come across some dubious scientific inference. What I am proposing to bracket are questions about the warrant of individual scientific inferences. This is part and parcel of my granular approach to compositional abduction.

To this point of the section, I have been suggesting that by focusing on abduction, rather than warranted abduction, philosophers of science might understand individual scientific inferences about specific experimental

[9] Norton (2021, pp. 5–6).

results without the presupposition that, in those cases, scientists are making warranted inferences and that they are somehow entertaining rival hypotheses and eliminating them. As noted in Chapter 1, I am not doing epistemology or metaphysics. I am not trying to argue that scientific abductive arguments for compositional hypotheses are warranted or that they show us what really exists in the world. Instead, I am describing what scientists are doing. Perhaps once there is an inventory of scientific practices, then philosophers of science can turn to the task of determining which are justificatory and which are not.

8.3 Explanatory Virtues

To this point in my discussion of Harman's views, I have indicated why I do not assume, going into any individual scientific interpretation of an experimental result, that the scientist will examine any rival hypotheses. Now, let me turn to the matter of how scientists cope with rival hypotheses once they have been introduced. Having introduced the need to eliminate rival hypotheses, Harman offers two sentences of guidance about how this might be done:

> There is, of course, a problem about how one is to judge that one hypothesis is sufficiently better than another hypothesis. Presumably such a judgment will be based on considerations such as which hypothesis is simpler, which is more plausible, which explains more, which is less ad hoc, and so forth. (Harman, 1965, p. 89)

Clearly, there is a wide range of options for developing this proposal. Harman does not say what simplicity is, what plausibility is, etc. He does not have an account of what it is for one hypothesis to explain more than another; he does not have an account of what is sometimes called "scope" or "breadth." Nor is his list of considerations meant to be exhaustive. There might be different weightings of the relative importance of items on the list. The weightings might be dynamic, changing from context to context. There might be individual scientific variations in the weightings, wherein one scientist will assign one set of weightings, whereas another will assign another set of weightings. Harman clearly left a lot of room to develop this story.

Given the open-ended character of IBE, it is highly likely that some version of it can be made to cover the cases I describe. What I have called "abduction" might be a special case of IBE in which there is only one hypothesis under consideration. In Chapter 5, I argued, in brief, that Hodgkin and Huxley took the membrane hypothesis to be unable to explain the overshoot of the action potential, whereas postulating sodium and potassium fluxes could explain the overshoot. One might get the IBE model to fit the case by proposing that Hodgkin and Huxley put low weights on simplicity, plausibility, and being non-ad hoc, but high weight on the greater scope of the sodium and potassium hypotheses. In Chapter 6, I noted, in brief, that Schiller and Carvey took the RGC theory of the Hermann grid illusion to explain the standard version of the illusion, but to be unable to explain the weakening of the illusion with a 45° rotation. By contrast, they took their S1 theory to be able to explain both the standard and rotated versions of the illusion. As with the Hodgkin and Huxley case, one might get the IBE model to fit by proposing that Schiller and Carvey put low weights on simplicity, plausibility, and being non-ad hoc, but high weight on the greater scope. So, it would likely be a mistake to try to show that IBE is wrong.

Of course, it is one thing to not be wrong. It is another to be illuminating. While IBE may not be refuted by actual cases, it does not help to illuminate the cases. Simply starting with Harman's IBE sketch, one would not have any idea how to apply it to the scientific study of the action potential or the Hermann grid illusion. To understand how IBE might apply to episodes in the history of science, the natural thing to do is examine the history of science.

8.4 IBE is Used in Science

To this point in the chapter, I have examined IBE as a possible model of how scientists evaluate the results of a single scientific experiment. Yet, that application is not a foregone conclusion. Let me return again to Harman's paper. In an effort to motivate the significance of the idea of IBE, Harman proposes that IBE is used in detective work, knowledge by testimony, and in scientific investigation. He comments that "When a scientist infers the existence of atoms and sub-atomic particles, he is inferring the truth of an explanation for various data which he wishes to account for" (Harman, 1965, p. 89). This passing comment has invited some philosophers, such as Lipton and Schurz, to treat IBE as a psychological hypothesis or as a description of how scientists make warranted inferences in particular cases, such as the evaluation of the results of individual scientific experiments. Yet, this is not the only application.

One might, instead, treat IBE as an account of a scientific community's theory choice. The rough and ready idea here is that it is one thing for an individual to claim that evolution by natural selection provides the best explanation of the blind cave forms in Kentucky resembling the nearby surface forms, but another thing to claim the biological community concluded that evolution by natural selection provides the best explanation of terrestrial biogeography. Very roughly, there is a distinction between explaining some individual fact and explaining some non-singleton set of facts.

In fact, in the literature, there is often a slide between introductions of IBE as a model for the explanation of some singular event and using it as a model of how scientists choose theories. Kevin McCain gives the following illustration of IBE: "You come home ready to eat the noodles that you saw in the refrigerator this morning, but they're gone. You only have one roommate, and he has been home all day. You infer that he ate the noodles" (McCain, 2019). Over the course of a few pages, McCain shifts to IBE as an account of the confirmation of the theory of natural selection and the oxygen theory of combustion.[10] These two applications of IBE should not be assumed to be identical.

My science-first approach and its resulting account of singular compositional abduction is not intended to be a rival to an IBE account of theory choice. Nevertheless, it might be useful input to such an account. One might think that singular compositional abductive inferences support hypotheses of some theory, hence be among the factors that influence scientific theory choice. Or, to put the matter another way, theory choice might be a function of singular compositional abductive inferences. Maybe singular compositional abduction would help articulate Harman's idea that theory choice might be a function of what it is for a theory to explain more. Maybe it would help articulate the idea of "explanatory scope" or "explanatory breadth."[11] For a concrete illustration, consider Thagard's 1978 account of theory choice. Thagard proposes three main criteria: consilience, simplicity, and analogy. For present purposes, I focus on consilience, where my approach may have the most to contribute.

Thagard uses the following notation for his account. Let a theory T be a set of hypotheses $\{H_1, H_2, \ldots, H_m\}$, A be a set of auxiliary hypotheses $\{A_1, A_2, \ldots, A_n\}$, C be a set of accepted conditions $\{C_1, C_2, \ldots, C_j\}$, and F be a set of facts $\{F_1, F_2, \ldots, F_k\}$ with $k \geq 2$.[12] T is consilient if, and only if, $T \cup A \cup C$ explains a non-singleton subset of F. Next let FT_i be the set

[10] See Lipton (2003, p. 56) for another example.
[11] Thagard (1978). Beebe (2009), among many others, also refers to explanatory scope.
[12] Strictly speaking, Thagard proposes that F is a set of classes of facts, although he does not specify what a class of facts is. This detail is inessential for my discussion.

of facts explained by T_i. This notation enables Thagard to formulate two concepts of consilience: (1) T_1 is more consilient than T_2 if, and only if, the cardinality of FT_1 is greater than the cardinality of FT_2 and (2) T_1 is more consilient than T_2 if, and only if, FT_1 is a proper subset of FT_2.

Before turning to the contributions of my account of compositional abduction, consider a general challenge for Thagard's account, namely, that his two conceptions of consilience do not capture some important features of scientific practice. The first conception of consilience does not capture the idea that theory choice typically involves a choice between two "competing" theories, that is, two theories that are concerned with at least some of the same facts. Let F be the union of the set of facts explained by, say, the Hodgkin–Huxley theory of the action potential and the set of facts explained by, say, the RGC theory of the Hermann grid illusion. In the event that these two sets differ in cardinality, Thagard's first conception will judge the Hodgkin–Huxley theory more consilient than the RGC theory, or *vice versa*. However, scientists do not often make theory choices like this. The choice is not between the Hodgkin–Huxley theory and the RGC theory. Instead, the target of scientific choices is more apt to be between Bernstein's membrane theory and the Hodgkin–Huxley theory or between the RGC theory and the S1 theory of the Hermann grid illusion. So, the first version of consilience suggests too many theory choices.

The second version, however, may not capture some consilience-based theory choices. Consider a case where FT_1 is a not proper subset of FT_2 and FT_2 is not a proper subset of FT but FT_1 and FT_2 have common members. Scientists might say of such a case that T_1 is more consilient than T_2 or that T_2 is more consilient than T_1, but Thagard's second version of consilience does not deliver an account of that scientific judgment. Thagard's second version only applies in cases where one theory explains everything another theory explains and more. This second version, thus, suggests too few theory choices.

Now consider three positive contributions my theory of compositional abduction may make to Thagard's theory. First, there is a striking gap in Thagard's theory: there is no account of what it is for $T \cup A \cup C$ to explain a fact. The outline of singular compositional explanation in Chapters 2 and 3 may not cover everything one wants in a theory of explanation, but there is an account of at least some explanations. Second, with a theory of singular compositional abduction in hand, it is easier to see that an abductive inference stemming from the explanation of a single fact or result is not the same as Thagard's inference based on consilience. Clearly the inference based on consilience must involve a consideration of the cardinality of two sets and the greater than relation. Singular compositional abduction need not consider these. Thagard's inference from consilience may be a "neighbor" of singular compositional abductive inferences in the sense that it may be used in conjunction with singular compositional abductive inferences, but it is not to be identified with singular compositional abduction. Thagard's inference from consilience is an "amalgamative inference" that involves considerations of multiple factors, such as what is explanatory and the cardinality of sets of explanations. Third, my account of singular compositional abduction includes an account of disconfirmation. Surely, Thagard's account of scientific theory choice should have something to say about the role of failures to explain. Thus, there are features of my compositional abductive account that would be useful additions to Thagard's view of theory choice.

8.5 Peircean Abduction and Inference to the Best Explanation

As a final point to round out the discussion of Harman (1965), I will touch on the connection between Peircean abduction and Harman's IBE. Harman begins the first section of his paper with a comment to orient his reader: "'The inference to the best explanation' corresponds approximately to what others have called 'abduction,' 'the method of hypothesis,' 'hypothetic inference,' 'the method of elimination,' 'eliminative induction,' and 'theoretical inference'" (Harman, 1965, pp. 88–89). As I read this, it is not a serious bit of Peirce scholarship. It is merely a bit of stage setting.

This has not, however, kept Peirce scholars from uncharitable readings of Harman. In "How did abduction get confused with inference to the best explanation?," McAuliffe lays the blame on Harman.

> Harman equates IBE with several terms that Peirce used for abduction: "'The inference to the best explanation' corresponds approximately to what others have called 'abduction,' 'the method of hypothesis,' 'hypothetic inference,' 'the method of elimination,' 'eliminative induction,' and 'theoretical induction.'" (Ibid., pp. 88–89) Harman does not refer to Peirce or any other author. There is no way to determine whether Harman is referring to Peirce's earlier writings, or to Peirce's later writings, or not referring to Peirce at all. (McAuliffe, 2015, p. 306)

Part of what makes this uncharitable is that Harman clearly does not equate IBE with abduction. He says IBE "corresponds approximately" to abduction. Another part is that it ignores the context of Harman's comment, namely, a study of the relationship between warranted enumerative induction and IBE, rather than a study of the history of philosophy. It is, of course, unclear what Harman means by "corresponds approximately," but the qualification does provide cover for Harman's orienting comment – a comment not meant to be a serious treatment of the history of philosophy. Yet a third reason the treatment is uncharitable is that it does not explore an obvious conjecture that might underlie Harman's thought about the approximate correspondence between IBE and abduction. Peirce claimed that abduction is "the operation of adopting an explanatory hypothesis" (Peirce, 1992, p. 231). Why is it not reasonable to think that the operation of adopting an explanatory hypothesis "corresponds approximately to" IBE? A charitable interpretation of Harman's text should attend to this.

McAuliffe also claims that "Contemporary philosophers of science have falsely cited Peirce's idea of abduction as a conceptual precursor to the modern notion of inference to the best explanation" (McAuliffe, 2015, p. 300).[13] McAuliffe does not spell out why this is false. What reason is there to think that Peirce's idea of abduction is not a conceptual precursor to IBE? After all, "being a precursor to" – like "corresponding approximately" – is weaker than being identical to. Moreover, what reason is there to think that contemporary philosophers of science intend to make any claims about what Peirce believed about abduction? Contemporary philosophers of science might have an agenda that differs from the agenda of Peirce scholars.

Finally, McAuliffe claims that "Peirce defined 'abduction' and then philosophers of science took the word, assigned it a new meaning, and used Peirce's work to give IBE the illusion of a respectable pedigree" (McAuliffe, 2015, p. 310). Again, this is a surprisingly uncharitable interpretation. As McAuliffe notes, Harman (1965) does not refer to Peirce. Lipton (2003) does not discuss Peirce's view; it only refers to his collected papers. Maybe many of the references to Peirce in the IBE literature are simply nods to his philosophical contribution, rather than attempts to draw on Peircean prestige.

8.6 Summary

My goal in this chapter has not been to refute Harman's conception of confirmation. A better take is that the goal has been to show how what

[13] Cf., Campos (2011, p. 419).

Harman said about IBE is compatible with my project. It is quite clear that his philosophical project is orthogonal to mine, so there need be no conflict. Harman was primarily concerned with the thesis that warranted enumerative induction is a special case of IBE, whereas I am concerned with the scientific interpretation of some experimental results. To this end, I have developed a theory of singular compositional abduction and reviewed its use in two cases studies. One should not assume that the IBE model Harman developed for his project is suited for my project. The key element here is that, in the interpretation of scientific experimental results, one should not assume that scientists take the confirmation of a compositional hypothesis to lie in the interpretation of a single scientific experiment.

CHAPTER 9

Lipton and Inference to the Best Explanation

Harman commented that, "When a scientist infers the existence of atoms and sub-atomic particles, he is inferring the truth of an explanation for various data which he wishes to account for." Since then, philosophers interested in IBE have frequently proposed that the use of IBE is pervasive in both everyday life and in science.[1] Despite the familiarity of this claim, there is typically little attention to details about how IBE relates to scientists, experiments, or journal articles. So, in theory, one can say that IBE is used in science without being committed to the details of IBE's relation to the scientific interpretation of the results of controlled experiments. Lipton does not take this path. He proposes that IBE is pervasive, but also commits to the applicability of his account to the scientific interpretation of experimental results. Section 9.1 will review Lipton's aspirations for his account of IBE.

Lipton's account of the scientific interpretation of the results of controlled experiments draws together views on contrastive explanations, Mill's Method of Difference, and controlled experiments. In using contrastive questions, Lipton draws on a tradition often traced to Bas van Fraassen's contention that explanation-seeking why-questions have the normal form, "Why P, but rather than Q?" Lipton links such questions to Mill's Method of Difference. Finally, Lipton proposes to link this to a controlled experiment in which a scientist asks, "Why is the mortality rate high in Clinic 1, rather than low?" Although many of the issues Lipton discussed have disappeared from more recent treatments of IBE, I review them here because they may serve as a useful foil for my views. At the very least, they may dull the suggestion that my account of the scientific interpretation of controlled experiments merely imports Lipton's account. Section 9.2 will review this.

On Lipton's version of IBE, scientists select the likeliest possible explanation of some phenomenon or phenomena based on which possible

[1] See, for example, Van Fraassen (1978), Lipton (2003), Bird (2005), McMullin (2013), Douven (2017a), McCain and Poston (2017b), and Park (2022).

explanation is the loveliest.[2] By the loveliest explanation, he means the explanation that provides the most understanding. By the explanation that provides the most understanding, he means the one that ranks best in terms of the explanatory virtues of mechanism specification, precision, scope, simplicity, fertility, and fit with background beliefs. One shortcoming of Lipton's discussion is that he essentially rehearses familiar ideas in the literature, rather than investigating how those familiar ideas are applicable to the scientific interpretation of an experimental result. Section 9.3 will review this.

9.1 The Scope of Lipton's IBE

In Chapter 1, I described the target of my theory: the scientific interpretation of certain experimental results given in the primary experimental literature. There I distinguished this project from that of describing the psychology of individual scientists as might be in play in the inferences scientists make in the laboratory. In the discussion of Harman (1965), I also distinguished this project from a project of theory choice, as illustrated in Chapter 8 with the example from Thagard (1978). Lipton (2003), however, has more sweeping aspirations for his theory of IBE than I have for my theory of singular compositional abduction. At times, his IBE model is intended to be an account of human psychology. At other times, it is a model for the scientific interpretation of individual experimental results. At still other times, it is a model of theory choice. Consider how these applications appear in his discussion.

In the introduction to his book, Lipton comments, "it is one thing to be good at doing something, quite another to understand how it is done or why it is done so well. It is easy to ride a bicycle, but hard to describe how it is done" (Lipton, 2003, p. 1).[3] Lipton's point is that it is hard to provide a philosophical characterization of IBE, because the mechanisms of IBE are part of a subconscious, maybe innate, human psychological endowment. In this regard, Lipton's view may be a precursor to Schurz's beachcomber story. In one particularly striking moment, Lipton comments, "We want to determine how illuminating that account is as a partial description of the mechanism inside the cognitive black box that governs our inductive practices" (Lipton, 2003, p. 71).[4]

[2] Lipton (2003, p. 60). [3] Cf., Lipton (2003, p. 12).
[4] See also Lipton (2003, pp. 19, 28, 61, 105, 109, 119f).

Maybe an infant's acquiring a grammar from primary linguistic data is in some respects like what Hodgkin and Huxley are doing in their interpretation of their experimental results in the experimental literature. Then again, it may differ in detail. Suppose that it is part of an infant's innate biological endowment to assume that it will be exposed to a natural language that contains nouns. In that case, the infant may never have to go through a process of weighing evidence to decide that the best explanation of the primary linguistic data is that there is a language containing nouns. Instead, the infant is "hard-wired" never to entertain the hypothesis that it is being exposed to a language without nouns. That might be labeled "IBE," but historians and philosophers of science should not simply assume that Hodgkin and Huxley are as innately "close-minded" in their thinking about the sodium hypothesis as are the infants in acquiring their natural language. Hodgkin was, for a time, resistant to the sodium hypothesis, but he did come around in time.

Although Lipton sometimes treats IBE as a psychological model, the preponderance of his discussion appears to be directed to what I mean by the scientific interpretation of the results of controlled scientific experiments. In chapter 3 of his book, he develops a theory of contrastive explanation that, in his chapter 5, he applies to Semmelweis's reasoning regarding childbed fever. A contrastive question has a fact-foil structure, "Why P, but not Q?" or "Why P, rather than Q?" In that chapter, Lipton relates such questions to controlled experiments by asking, "Why did the women in clinic 1 have a high mortality rate, whereas the women in clinic 2 did not?" This is the topic I will take up in Section 9.2.

Despite the attention Lipton pays to the application of IBE to controlled experiments in Lipton (2003), Lipton (1998) contains some comments that resist that interpretation. In an exchange that has received little attention in the abduction/IBE literature, Suppe (1998a) argued that IBE does not provide a good account of the structure of a scientific paper. In reply, Lipton (1998) comments that,

> First, the core idea that scientists argue for hypotheses by revealing their explanatory power does not require the consideration of alternatives, and in some cases we should not expect articulated alternatives to be on offer. ... Second, even when there is more than one hypothesis on offer, we should not expect them all to appear in a single paper: different scientists may press the case for different hypotheses, without considering the opposition. (Lipton, 1998, p. 408)

I think that Lipton is exactly right on these two points. His reply is just what I was noting in my reply to Schurz's beachcomber example. Lipton recognizes that, at a given moment in time, a scientist may only entertain one hypothesis. Further, he recognizes that the confirmation of some hypothesis over rivals might only emerge over a period of extended scientific investigation.[5] I also think that, in his rejoinder, Suppe aptly comments that, "This is a surprising move" (Suppe, 1998b, p. 423). The comment is apt since there is little in the rest of Lipton's work that would have suggested his reply.

Finally, Lipton suggests the possibility of applying IBE to theory choice when it comes to issues of the putative incommensurability of competing theories and the interpretation of so-called crucial experiments.[6] In a nutshell, a crucial experiment is one in which theory T_1 explains some result that theory T_2 does not. Incommensurability, however, threatens to block this interpretation, since the different theories have different results. Incommensurability proposes that T_1 explains one set of results, whereas T_2 explains a different, nonoverlapping set of results. Thus, both theories explain results that the other does not, so there is no one experimental result to serve as a crucial experiment. Lipton's proposed resolution is to look at T_1's explanation of its data and T_2's explanation of its data, and then pick the theory that offers the best explanation of its results.[7] Thus, Lipton has IBE applied to theory choice.

As suggested in Chapter 1 and in my discussion of Harman, I think it is important to critically examine the different possible applications of an IBE model. It might be a model of innate human psychology, or a model of the conscious interpretation of individual experimental results, or a model of theory choice. The bare idea of IBE is open-ended enough to apply to all these cases, but philosophers of science should not assume that one and the same version of IBE will apply in each of these cases. In what follows, I will examine Lipton's proposal for the application of IBE to the explanation of experimental results.

9.2 Causal Explanation, Contrastive Questions, and Controlled Experiments

In Chapter 3, there was an invitation to consider an explanation-seeking why-question:

[5] See also Lipton (2003, p. 90). [6] Lipton (2003, pp. 68–69).

[7] It is a surprising coincidence that Lipton's account here, like Thagard's in the last chapter, invites choosing the Hodgkin–Huxley model over the RGC theory (or vice versa), because the Hodgkin–Huxley model offers a better explanation of its phenomena than the RGC theory offers of its phenomena (or vice versa).

(HH) Why did axon no. 15 *in seawater* display an initial *inward* current when depolarized by 65 mV, but *in a sodium-free medium* display an initial *outward* current when depolarized by 65 mV?

In that chapter, I called (HH) a "controlled experiment question." Referring back to the Y-maze experiment by Spence and Lippitt, one might also recognize a similar question.

SL (Spence-Lippitt): Why did the hungry rat go to the arm with the food, whereas the thirsty rat went to the arm with the water?

I proposed to recognize questions such as (HH) and (SL), because doing so helped articulate the reasoning the scientists gave in support of compositional hypotheses. I proposed that Hodgkin and Huxley and Tolman (who relied on the Spence–Lippitt experiment) were engaged in compositional abduction in the context of a controlled experiment.

Lipton (2003) provides a similar account. Lipton proposes that IBE be developed around a causal model of explanation: to explain an event is to give some information about its causal history.[8] But what information about an event's causal history provides a good explanation? Answer: Information that meets our interests. What are those? They are interests specified by contrastive questions, such as "Why did Adam eat the apple, rather than the pear?"[9] Further, Lipton proposes that contrastive questions can be linked to the interpretation of the results of controlled experiments.

Despite the similarities between my account and Lipton's, there are four significant differences. First, Lipton's proposal derives from the debate over contrastivism in the theory of explanation.[10] Mine does not. Second, origins aside, there are differences between contrastive questions as understood in the contrastivism debate and controlled experiment questions as I have described them. Third, although there are instances of contrastive questions that reasonably pair with controlled experiments, there are also instances where they do not. Fourth, following Lipton, one might suspect that there is a relationship between controlled experiments and a "method of compositional differences." This latter method is not any of Mill's methods but may be inspired by Mill's method of difference. Although Lipton might suggest a connection between controlled experiments and a method of compositional difference, there are clear cases where compositional controlled experiments do not fit the method.

Contrastivism. Contrastive questions, answered by contrastive explanations, are sometimes framed as a response to Sylvain Bromberger's theory of

[8] Lipton (2003, p. 30f). [9] Lipton (2003, p. 33f). [10] For a recent review, see Shan (2019).

explanation in terms of correct answers to why-questions.[11] For this, Bromberger proposed a normal form of why-questions. To a first approximation, such questions begin with "why" and contain an inner yes-no question, that is, a question whose answer is either yes or no. Thus, "Why does copper turn green when exposed to air?" is a why-question in normal form. The yes-no question is, "Does copper turn green when exposed to air?"

Bas van Fraassen was an early supporter of an alternative theory of the normal form of why-questions.[12] According to van Fraassen, the proper form of the question,

1) Why did Adam eat the apple?

is something like one of the following:

1a) Why was it *Adam* who ate the apple?
1b) Why was it the *apple* Adam ate?
1c) Why did Adam *eat* the apple? (Van Fraassen, 1980, p. 127)[13]

1a)–1c) are, thus, supposed to disambiguate what is meant by 1). Alternatively, 1a)–1c) are supposed to make explicit what a speaker implicitly means by 1). As van Fraassen states it, "The correct, general underlying structure of a why-question is therefore 'Why (is it the case that) P in contrast to (other members of) X?' where X, the contrast-class, is a set of alternatives" (Van Fraassen, 1980, p. 127). Subsequent work has sometimes glossed 1a)–1c) along the following lines:

1a') Why was it Adam, rather than Brad, who ate the apple?
1b') Why was it the apple, rather than the pear, Adam ate?
1c') Why did Adam eat, rather than discard, the apple?

or perhaps

1a") Why did Adam eat the apple and Brad did not?
1b") Why did Adam eat the apple and not the pear?
1c") Why did Adam eat the apple and not discard it?

Sometimes these sentences are described in terms of facts and foils. Adam eating the apple is the fact, whereas Brad eating the apple, Adam

[11] Bromberger (1966) is the *locus classicus* for this theory. It is an especially careful and detailed discussion, but for present purposes, I will omit many of the details that are tangential to my concerns. In this regard, my discussion follows some of the important contributions to the contrastivism debate. See, for example, Van Fraassen (1980) and Temple (1988).

[12] Although discussions of contrastivism frequently refer to Van Fraassen (1980), this work draws upon what was, at the time, unpublished work by Jon Dorling, Alan Garfinkle, and Bengt Hansson. Garfinkle's work subsequently appeared as Garfinkel (1982).

[13] See also Temple (1988, p. 142) and Shan (2019, p. 2).

eating the pear, and Adam discarding the apple are foils.[14] The fact actually occurred, but the foils did not.[15]

While there are many points at which my project differs from the contrastivist project, let me mention just two before delving into further issues surrounding contrastivism. First, I note that, contrary to a long tradition in the philosophy of science, I do not take all explanations to be answers to why-questions. Some singular compositional explanations are answers to why-questions, but others are not. One might ask, "Why did this action potential overshoot zero?" This might be understood as a request for an account of how lower level processes bring about this process. This would be a request for a singular dynamic compositional explanation. One might ask, "Why is this axonal membrane capacitance 0.9 $\mu F/cm^2$?" This would be a request for a singular standing compositional explanation. Both of these questions are types of explanation-seeking why-questions, but there are other explanation-seeking questions. So, one might ask, "What is water made of?" Answer: Molecules of H_2O, OH^- ions, H_3O^+ ions, and H_2O oligomers. One might ask, "What are human retinas made of?" in the sense of "What cells make up human retinas?" Answer: Rods, cones, amacrine, bipolar, and horizontal cells. These would be requests for singular analytic compositional explanations. To understand explanation through the lens of why-questions is too narrow to do justice to scientific practice.

A second respect in which my project differs from van Fraassen's contrastivist project is that, even if one recognizes a wider range of questions with which to seek explanations, it is orthogonal to my concerns to determine whether there is, or is not, some canonical form for these questions. Maybe Bromberger's or van Fraassen's account of the normal form of explanatory why-questions is correct and maybe one could develop a normal form for compositional questions. Although that may be a worthy philosophical project, it is not my project.

Contrastive Questions versus Controlled Experiment Questions. Although my project differs from that of van Fraassen and his successors, it could, of course, happen that controlled experiment questions are just contrastive questions in van Fraassen's sense. So, let me point out three differences between "van Fraassen style" contrastive why-questions and compositional controlled experiment why-questions.

Consider, first, the fact that there is supposed to be a semantic or pragmatic relation between 1) and 1a)–1c). Whether the relation is

[14] See, for example, Carroll (1997) and Lipton (2003). [15] Lipton (2003, p. 34).

semantic or pragmatic does not matter for my present point. 1a)–1c) are disambiguations of 1); 1a)–1c) make explicit what is implicit in 1). By contrast, the HH and SL questions are not disambiguations. The two conjuncts in the HH and SL questions relate to two empirically distinct explananda.

A second feature of contrastive questions, as commonly understood, is that the fact is supposed to have occurred, whereas the foil has not.[16] Adam ate the apple, but Brad did not eat the apple, Adam did not eat the pear, and Adam did not discard the apple. This central presupposition is incorporated into, for example, David-Hillel Ruben's treatment of contrastive explanation. Ruben (1987, pp. 35–36) proposes that explaining why Adam, rather than Brad, ate the apple, is a matter of explaining Adam's eating the apple and explaining how Adam's eating the apple "eclipses" (a more general concept than "causally prevents") Brad's eating the apple. In this case, the thought behind the eclipsing might be that if Adam has eaten the apple, there is nothing left of the apple for Brad to eat.[17] By contrast, in both the HH and SL questions, both conjuncts occurred. In the HH case, neuron no. 15 displayed an initial inward current when depolarized by 65 mV in a sodium-containing medium and it later displayed an initial outward current when depolarized by 65 mV in a sodium-free medium. In the SL case, one hungry rat went to the arm with the food, whereas another thirsty rat went to the arm with the water.

Of course, one might observe that, in the HH case, the axon's displaying an initial outward current means that it did not display an initial inward current. This might lead one to another explanation-seeking why-question,

HH': Why did the axon display an initial inward current when depolarized by 65 mV in a sodium-containing medium, but did not display an initial inward current when depolarized by 65 mV in a sodium-free medium?

Similarly, one might observe that, in the SL case, the thirsty rat's going to the arm with the water means that it did not go to the arm with the food. This might lead one to the question,

SL': Why did the thirsty rat go to the arm with the water, but not to the arm with the food?

[16] Temple (1988, p. 144), Lipton (2003, p. 34).
[17] Lipton (1987) presents three counterexamples to Ruben's analysis.

Moreover, it is perfectly reasonable to suppose that the HH experiment could prompt HH' and the SL experiment could prompt SL'. Finally, HH' and SL' might count as instances of the more familiar van Fraassen-style contrastive why-question. So, one might think that there is, after all, a connection between controlled experiment questions and contrastive questions.

All of the points in the last paragraph can be conceded. Controlled experiments may raise many questions. Some of these may be familiar contrastive questions. My point, however, is that controlled experiments also raise some questions that are not the familiar van Fraassen-style contrastive questions. Nothing in the last paragraph undermines the idea that HH and SL are not familiar contrastive why-questions. HH and HH' are different questions; SL and SL' are different questions. The second conjunct of HH contains a stronger presupposition than does the second conjunct of HH'. HH presupposes that there is an initial outward current when the axon is depolarized by 65 mV in a sodium-free solution, but HH' does not. HH is inconsistent with the axon displaying no current when the axon is depolarized by 65 mV in a sodium-free solution, whereas HH' is consistent with this. Similarly, *mutatis mutandis* for SL. For this response to work, one must recognize that there is no such thing as *the* question a controlled experiment may prompt.

A third feature of the contrastive questions, as often understood, is that they share what might loosely be described as common arguments, while differing in a single argument. So, for instance, Ia') and 1a") have the apple and eating as common arguments but differ in Adam and Brad. 1b') and 1b") have Adam and eating as common arguments but differ in the apple and the pear. Controlled experiment questions need not have this limitation. Hodgkin and Huxley voltage clamped neuron no. 15 first in a sodium-containing medium, and then in a sodium-free medium, and then in a sodium-containing medium, each time measuring the resulting currents. This experiment has neuron no. 15 as the common argument with the medium being the difference. But the two conjuncts of the HH question differ not just in one argument, but in two. The first conjunct asks why the axon displayed an initial *inward* current when depolarized by 65 mV in a *sodium-containing* medium and the second asks why an initial *outward current* when depolarized by 65 mV in a *sodium-free medium*? The direction of the current and the contents of the medium varied. The situation with the Spence and Lippitt experiment is the same. The hungry rat familiarized with the maze went to the arm with the food; the thirsty rat familiarized with the maze went to the arm with the water. The familiarity with the maze was a common feature of both rats, but one was hungry and the other thirsty and one went one direction and the other went another.

It should be noted that in many discussions of contrastive questions, facts and foils are represented using sentential letters, such as P and Q.[18] For many purposes, this simple notation is perfectly adequate. Nevertheless, one unfortunate consequence of it is that it may obscure some of the structure implicit in contrastive questions. It may even lead some to suppose that I am misinterpreting the contrastive view. Fortunately, I am not alone in drawing attention to this implicit structure. Thomas Grimes uses differences among what I am calling "arguments" to distinguish different contrasts.[19] Taking a cue from Grimes, one might say that 1a') is a request for an explanation of an "S-fact" because it concerns the subject of the sentence; 1b') is a request for an explanation of an "O-fact" because it concerns the direct object of the sentence; and 1c') is a request for an explanation of an "A-fact" because it concerns the activity instance of the sentence. Those who consult Grimes (1993) will find that I am not simply taking over Grimes's scheme.[20] I am, instead, merely drawing attention to its recognition of the argument structure of contrastive questions. That is why I claim to be "taking a cue" from Grimes.

To this point, I have often qualified my claims by noting that contrastive questions *as commonly understood* have such and such features that my controlled experiment questions do not. I have also qualified my claim by referring to *van Fraassen-style* contrastive questions. Nothing, however, prevents a philosopher of science from abandoning earlier presuppositions about what counts as a contrastive question. Do that. Let controlled experiment questions be contrastive questions. For my purposes, it is enough that I have a correct account of what controlled experiments questions are. Whether they are, or are not, ultimately counted as contrastive questions is a tangential matter.

Lipton on Contrastive Questions and Controlled Experiments. I have proposed that controlled experiments raise what I have called controlled experiment questions. Hodgkin and Huxley's first experiment in Hodgkin and Huxley (1952a) provokes the HH question, "Why did the axon display an initial inward current when depolarized by 65 mV in a sodium-containing medium, but an initial outward current when depolarized by 65 mV in a sodium-free medium?" Similarly, Lipton proposed that controlled experiments raised van Fraassen-style contrastive questions,

[18] See, for example, Ruben (1987) and Temple (1988).
[19] Grimes (1993).
[20] Grimes (1993, p. 32).

such as "Why did Adam eat the apple, but not the pear?" A quick read of Lipton (1987, 1990, 1991) reveals that Lipton was quite familiar with van Fraassen-style contrastive questions. Moreover, Lipton is quite explicit in thinking that this analysis applies to Semmelweis's reasoning regarding childbed fever: "This case is a gold mine for inferences to the best contrastive explanation" (Lipton, 2003, p. 75). I now want to argue that Lipton is mistaken in aligning van Fraassen-style contrastive questions with controlled experiments. Lipton is right that Semmelweis relied upon controlled experiments in his efforts to determine the cause of childbed fever. Where he goes wrong, however, is in thinking that the reasoning behind controlled experiments is generally captured by van Fraassen-style contrastive questions/explanations. The analysis works sometimes, but it fails at others.

To begin with, let me show a case where it appears to work. One of the central observations for Semmelweis was the mortality rate due to childbed fever was higher in clinic 1 than in clinic 2. To deploy the van Fraassen-style contrastive question, let us begin with

(2) Why was the mortality rate high in clinic 1?

This might be disambiguated as follows,

(2a) Why was the mortality rate high in clinic 1 and not low in clinic 1?
(2b) Why was the mortality rate high in clinic 1 and not in clinic 2?

Let us assume that the relevant disambiguation is (2b), that (2b) gives an explanation-seeking why-question that concerned Semmelweis. Lipton proposes that atmospheric conditions would not explain why the mortality rate was high in clinic 1 and not in clinic 2, since atmospheric conditions were the same in both of the clinics. This example works as far as it goes.

A weak point in Lipton's analysis, however, is that scientists often make stronger commitments than are captured by van Fraassen-style contrastive questions. Semmelweis did not believe merely that the mortality rate was not high in clinic 2. The belief that the mortality rate was not high in clinic 2 is consistent with the belief that the mortality rate was low in clinic 2 and that the mortality rate was extremely high in clinic 2. Semmelweis clearly believed that it was low in clinic 2, or at least lower than in clinic 1. This belief shaped Semmelweis's experimental design. Semmelweis noted that many times a day a priest walked through clinic 1 ringing a bell and considered the idea that this might be the cause of the higher mortality rate in clinic 1. He, therefore, arranged to have the priest avoid passing through much of the clinic and to stop ringing the bell. This had no effect on the mortality rate. Notice that if Semmelweis believed that the mortality rate was higher in clinic 2 than in clinic 1, his experiment with the priest would have likely involved the priest going to *both* clinic 1 and clinic 2. That experiment would have the hopeful outcome of reducing the mortality rate in clinic 2. Clearly, Semmelweis's reasoning was guided by the belief that the first clinic had a higher mortality rate than the second clinic, rather than the belief that the first clinic had a different mortality rate than the second. One may, of course, propose that Semmelweis had both this first belief and the second belief, since the first belief logically entails the second. Nevertheless, it was clearly the first belief that drove Semmelweis to conduct the experiment he did. The point here reiterates the importance of the distinction between the HH and HH' questions and SL and SL' questions introduced above.

Controlled Experiments and a Method of Compositional Difference. Lipton (2003) proposed that his version of IBE was an improvement over Mill's method of difference for causal inference. Rappaport (1996) and Scholl (2013, 2015), however, have argued that Lipton's account does not differ from Mill's methods. Further, Norton (2021) embraces Scholl's analysis.[21] This exchange, thus, raises the question of the extent to which my account of singular compositional abduction is, or is not, merely an instance of a difference principle for compositional, as opposed to causal, inference. Let the difference principle be the following:

> If an instance in which the phenomenon under investigation occurs, and an instance in which it does not occur, have every circumstance save one in common, that one occurring only in the former; the circumstance in which alone the two instances differ, is a component of the phenomenon.

This difference principle is clearly inspired by John Stuart Mill's method of difference.[22] Here, I wish to emphasize that my term "inspired" is meant to leave open the extent to which Mill may or may not have accepted this difference principle. That is an issue I leave to specialists in Mill's philosophy of science. My concern is to argue that there are cases of singular compositional abduction that do not work according to a natural reading of this simple difference principle.

Recall that in Hodgkin and Huxley's first experiment they postulate that depolarizing axon no. 15 by 65mV increases the membrane's permeability to sodium. As a result, when there is excess extracellular sodium, sodium flows into the axon, whereas when there is no extracellular sodium,

[21] See also Barnes (1995). [22] Mill (1843, p. 455).

sodium flows out of the axon. Following the compositional difference principle, one should expect to find Hodgkin and Huxley postulating something that differs between the case when the medium contains sodium and the case when the medium lacks sodium. And there is such a difference: the difference in the direction of the flow of sodium. But Hodgkin and Huxley also postulate something in common to the two cases: both involve an increase in the membrane's permeability to sodium.

Next, recall that, in support of the cognitive map hypothesis, Tolman (1948) cited an experiment performed by Spence and Lippitt. Recall that, in this Y-maze experiment, the hungry rats went to the arm with food and the thirsty rats went to the arm with water. Tolman, thus, postulated a difference between the two cases, hence acted in accordance with the compositional method of difference. Nevertheless, he also postulated something common: both the hungry and the thirsty rats formed a cognitive map of the Y-maze. The upshot of these two examples, therefore, is that singular compositional abduction is not simply the application of a Mill-inspired method of compositional difference.

9.3 Understanding and the Explanatory Virtues

Although Lipton's view is often described as inference to the best explanation, the somewhat more detailed formulation of his view is that scientists use inference to the loveliest potential explanation. A potential explanation is in all respects an explanation, except for truth. Lovelier explanations contribute more understanding. Further, explanations that are lovelier are explanations that are more likely. Finally, the explanatory virtues of mechanism, precision, scope, simplicity, fertility, and fit with background belief contribute to understanding. In advancing these claims, Lipton is rehearsing familiar ideas.

While there is nothing intrinsically wrong with these familiar ideas, they do not contribute much to the theory of singular compositional abduction. Some are not worked out in sufficient detail. Others are not applicable to a singular compositional explanation. Work through Lipton's list of virtues in order.

Mechanism. In a singular dynamic compositional explanation, one might suppose that insofar as a potential explanation of S Ψ-ing does not specify the x_i's or the φ_I-ings, one has failed to specify a mechanism. Yet there are two striking cases of widely accepted compositional explanations in which scientists are unbothered by explanations of S Ψ-ing that do not specify the x_i's or the φ_i-ings.[23] In Chapter 5, I noted that Hodgkin and Huxley believed that an internal microelectrode that touched the wall of the axon damaged the axon. How did they arrive at that conclusion? They hypothesized that the wall was damaged by the touching, since that would explain the subsequent electrical measurements. They had no other means of inspecting the membrane to determine in detail what had happened to membrane. There is a clear sense in which Hodgkin and Huxley could not specify the mechanism of the damage, yet this apparently did little to affect the acceptance of their explanation of the effects of touching the axonal membrane. A second case involves "run down." Hodgkin and Huxley believed that, over the course of an experiment, the axons were running down. They were dying. Yet, Hodgkin and Huxley did not specify what parts of the cells were changing or what activity instances of those parts were changing. Instead, they simply postulated changes in unspecified activity instances in unspecified individuals as responsible for the changes in axonal responses. What are philosophers of science to make of these cases vis-à-vis a "virtue for mechanism?" This is worth exploring, but Lipton offers nothing to lead the way.

Precision. Lipton proposes that "we understand more when we can explain the quantitative features of a phenomenon, and not just its qualitative ones" (Lipton, 2003, p. 122). Given that Lipton does not explain what it means to "understand more," the phrase is probably vague enough to be able to resist refutation. Nevertheless, even if the account is not wrong, it is also not illuminating. To get the point, recall that Hodgkin and Huxley's "Quantitative Description" paper, the fifth in their 1952 series, provided a quantitative model of their theory of the action potential. What did this quantitative model do for Hodgkin and Huxley? Did it help them understand more or did it merely provide them with greater confirmation of hypotheses they already understood? Lipton's comments are of no help here.

Scope, simplicity, and fruitfulness. The remaining virtues, as Lipton develops them, are not features of individual singular compositional explanations. They are features of an explanation in a sense in which an explanation is an explanation of many phenomena. (This issue was discussed in the context of Thagard's treatment of the consilience of

[23] Chapter 6 noted that Hermann's explanation of the grid illusion failed to provide a mechanism in the same sense that it failed to specify x_i's and ϕ_I's. This case differs from the HH case insofar as the HH explanation has been widely accepted as unproblematic, but the evidence that Hermann's explanation was acceptable is harder to come by.

abductions in Chapter 8.) Take a concrete case. Lipton claims that "An explanation that explains more phenomena is for that reason a lovelier explanation" (Lipton, 2003, p. 122). That is a common enough formulation of what scope is, but it is a sense of "explanation" in which one explanation explains multiple phenomena, that is, multiple explananda. It is not the sense of "explanation" that is in play in the explanation of one of the initial inward currents of Hodgkin and Huxley's axon no. 15. The explanation in the Hodgkin–Huxley case is of a single spatiotemporal occurrence. It involves a specific set of sodium ions moving into the axon at a single time and place. The explanans and explananda do not recur. This is just a feature of the explanation being singular. To get a general notion of an explanation that explains many phenomena, one might say that there is the general hypothesis – maybe what Bogen and Woodward would count as a phenomenon – of sodium carrying the initial inward current of the action potential. This general hypothesis might be said to have the scope it does in virtue of, say, the number of specific individual explanations in which it occurs.

Simplicity might be understood in many different ways. Lipton recognizes this but describes a conception of simplicity understood as a feature of the way in which a single hypothesis might be applied in the explanation of many explananda: "Again, simplicity, in its various guises, contributes to loveliness. For example, some forms of simplicity enable us to achieve one of the cardinal goals of understanding, namely to reveal the unity that underlies the apparent diversity of the phenomena" (Lipton, 2003, p. 122). The unity in the apparent diversity lies in one and the same general hypothesis – a Bogen-Woodward phenomenon? – occurring among the explanantia of multiple explanations. And while philosophers of science might accept this as a *bona fide* explanatory virtue, it is not an explanatory virtue of a singular compositional explanation.

The same point applies *mutatis mutandis* to Lipton's idea of fruitfulness, as borrowed from Thomas Kuhn. "Fruitfulness, the power of a theory to, in Kuhn's words, 'disclose new phenomena or previously unnoted relationships among those already known' [(Kuhn, 1977, p. 322)], is linked to scope and simplicity, and is again an explanatory virtue" (Lipton, 2003, p. 122). Part of what Lipton and Kuhn are pointing to under the heading of fruitfulness is straightforwardly described as the power of some hypothesis to explain multiple new phenomena. This, again, is not a feature of a singular compositional explanation. Disclosing previously unnoted relationships among already known phenomena might be a matter of showing how one compositional hypothesis might explain two phenomena, thereby indicating that they are related, when this relation was not previously noted.

Another way of making the foregoing point about the scope, simplicity, and fruitfulness of a compositional explanation is that it presupposes an account of singular compositional explanations. To articulate what these virtues are, one wants an account of singular compositional explanation. Thus, just as a theory of singular compositional explanation might flesh out Thagard's account of the consilience of abductions, so it might flesh out Lipton's account of scope, simplicity, and fruitfulness.

Fit with the scientific background. Finally, there is what Lipton describes as an explanation's fit with the scientific "background." Chapter 4 provided a sketch of how pedestrian scientific beliefs play a role in compositional abductive arguments. Beliefs about the reliability of thermometers, the sodium concentration of seawater, and the existence of junction potentials are such pedestrian background beliefs. Lipton, however, has in mind something more global, namely, "background beliefs about what sorts of accounts are genuinely explanatory. For example, at given stages of science no appeal to action at a distance or to an irreducibly chance mechanism could count as an adequate explanation, whatever its empirical adequacy" (Lipton, 2003, p. 122). There is a lot to be said about this topic that is particularly germane to compositional abduction, although Lipton has nothing to say about it.

Consider some suggestive comments on this score. One of the more striking illustrations of a "Liptonian" background belief in cognitive science, neuroscience, and psychology concerns the legitimacy of mental representations and the legitimacy of postulating mental representations to explain behavior. Although the issue is not typically framed in these terms, skepticism about mental representations is often intertwined with skepticism about compositional abductive reasoning. One can see the interplay in B. F. Skinner's *Verbal Behavior.*[24]

> It has generally been assumed that to explain behavior, or any aspect of it, one must attribute it to events taking place inside the organism. In the field of verbal behavior this practice was once represented by the doctrine of the expression of ideas. An utterance was felt to be explained by setting forth the ideas which it expressed. If the speaker had had a different idea, he would have uttered different words or words in a different arrangement. If his utterance was unusual, it was because of the novelty or originality of

[24] Clearly, Skinner's book falls outside this book's focus on the primary experimental literature. Moreover, Skinner is self-conscious about this. See Skinner (1957, p. 11).

> his ideas. If it seemed empty, he must have lacked ideas or have been unable to put them into words. If he could not keep silent, it was because of the force of his ideas. If he spoke haltingly, it was because his ideas came slowly or were badly organized. And so on. All properties of verbal behavior seem to be thus accounted for. (Skinner, 1957, p. 5)

> The difficulty is that the ideas for which sounds are said to stand as signs cannot be independently observed. If we ask for evidence of their existence, we are likely to be given a restatement in other words; but a restatement is no closer to the idea than the original utterance. (Skinner, 1957, p. 6)

> There is obviously something suspicious in the ease with which we discover in a set of ideas precisely those properties needed to account for the behavior which expresses them. We evidently construct the ideas at will from the behavior to be explained. There is, of course, no real explanation. (Skinner, 1957, p. 6)

One of the principal features of compositional abduction, like abduction more generally, is that it is used as part of a case for the existence of things that are neither directly measured nor directly manipulated. This is what Skinner evidently has in mind in objections to what cannot be independently observed. The existence of a capacity current and ionic currents during Hodgkin and Huxley's voltage clamp experiments, the damage to an axon when a microelectrode touches the membrane, or the run down of an axon are all instances in which Hodgkin and Huxley postulated unseen goings on. And, when using abduction, the scientist does aim to construct hypotheses that are explanatory.

Verbal Behavior was, of course, published almost a decade after Tolman's "Cognitive Maps" paper. Nevertheless, Tolman was working within a broadly behaviorist context that was skeptical of his cognitive map hypothesis. To push back against behaviorist resistance, Tolman did not pursue a strategy of championing the explanatory value of compositional abduction. He did not pursue the kind of general strategy Skinner pursued in *Verbal Behavior*. Instead, he tried to provide experimental results that he thought could be explained by the cognitive map hypothesis but not by other stimulus–response hypotheses. In pursuing this strategy, he was evidently banking on his fellow psychologists sharing common ground on what counts as a good explanation in individual concrete cases. He was banking on his fellow psychologists recognizing the acceptability of compositional explanations invoking cognitive maps in concrete cases, even if, in the abstract, they rejected compositional explanations invoking the mental. This is an idea that probably merits attention as a case study in its own right.

So, Lipton is right that there is some philosophy of science story that should be told about background beliefs regarding abduction. Hodgkin and Huxley clearly shared any number of background beliefs with their fellow physiologists that informed their abductive reasoning. Moreover, Lipton, following tradition, is right that there is something one might informally or pre-theoretically describe as the background belief that compositional abduction is, or is not, an acceptable form of reasoning. How to describe this theoretically, when scientists may be thought to lack a concept of compositional abduction, is not trivial. However one theoretically works out the idea of the common ground that Hodgkin and Huxley shared with other physiologists, this theory leaves untouched large parts of the account of how Hodgkin and Huxley interpreted the results of their multiple controlled experiments in support of their theory of the action potential. Hodgkin and Huxley did not rely on the background they shared with other physiologists to carry the day in favor of their theory of the action potential. The crucial evidence came from the interpretation of specific experimental results. Articulating how they did that has been the goal of this book.

9.4 Summary

Any reader familiar with Lipton (2003) will know that there are large parts of Lipton's work that I have completely ignored. Lipton has chapters on Bayesian abduction, the Raven paradox, the putative connection between loveliness and truth, and the proposal that prediction of new data is more valuable than the explanation of old data. Lipton also devotes a lot of attention to arguing that the IBE account of confirmation is superior to the HD account. Many will also object that I have not devoted sufficient attention to replies that Lipton might give to my analyses.

Perhaps there is philosophical interest in a more complete examination of Lipton's book, but that is tangential to my concerns. I am trying to advance a science-first approach to the confirmation of compositional hypotheses. Further, in proposing that scientists sometimes use singular compositional abduction, I have a theory of singular compositional explanation and detailed scientific case studies to back it up. As this book prioritizes the development of the positive project, I have focused my attention on the features of Lipton's account that will best illuminate my own account and I have devoted less attention to an exhaustive historical review.

CHAPTER 10

Manipulationist New Mechanism

The Peircean and IBE traditions have had little interest in theories of explanation. The New Mechanists, for their part, focus on a theory of explanation, but have had only a limited interest in abduction or IBE. Rather than considering an abductive/IBE approach to confirmation, some New Mechanists – the manipulationist New Mechanists (MNM) – have proposed a "manipulationist" approach. Here, I will contrast my account of abductive confirmation with the latest iteration of this approach, namely, Carl Craver, Stuart Glennan, and Mark Povich's "matched inter-level experiments" (MIE) account.

In rough outline, this chapter has the following organization. MIE is meant to describe actual scientific practice, or it is not. If it is, then there are features of scientific practice that it often fails to capture. This opens the door to an abductive approach that captures those features. If MIE is not meant to describe actual scientific practice, then again this opens the door to an abductive approach that captures those features. Thus, for my purposes, it is immaterial whether MIE is meant to describe actual scientific practice or not.

Section 10.1 will provide a barebones introduction to the MIE account of interlevel experiments along with an imagined example. Section 10.2 will describe two alternatives to the MIE interpretation of the imagined interlevel experiments: HD confirmation and abductive confirmation. This section also further develops the idea of singular interlevel abduction as a contrast with singular compositional abduction. Section 10.3 will move from the imagined example to two actual instances of the scientific use of interlevel experiments. One is borrowed from Kaplan (2012), and the other from Prychitko (2021). One might think that these instances illustrate MIE in action in science, but in point of fact the scientists instead relied upon a combination of interlevel and intralevel experiments. The MIE approach does not allow this, whereas the abductive approach does. Section 10.4 will offer a diagnosis of why the manipulationist approach is unsuited to addressing the combination of interlevel and intralevel experiments found in actual practice. It will argue that there is no easy fix for MIE. Sections 10.3 and 10.4 examine MIE as a description of how scientists sometimes confirm compositional hypotheses. In contrast, Section 10.5 examines various proposals that align in claiming that MIE is not meant to describe actual scientific practice at all. For what it is worth, these strike me as dubious interpretations of GC&P's goals for MIE, but my official view here is noncommittal. What matters to me is that these approaches leave open the door to an abductive approach to the scientific confirmation of compositional hypotheses.

I might note, at the outset, that my presentation will be unlike the standard fare in the MNM literature. First of all, I will largely stick with my own terminology, rather than that offered by the New Mechanists. I do not think the differences make a difference, so there is no need for a reformulation. Second, although I will refer to Craver's earlier works at multiple points, I will not provide a serious account of the historical development of MIE. I will not review Craver's initial description of interlevel experiments in Craver (2002), nor the "mutual manipulability" version of the approach in Craver (2007), nor the subsequent critique of mutual manipulability centering on Baumgartner and Gebharter (2016). There is already an abundance of writing on that.[1] Finally, I will not be concerned with the details of ideal interventions. There is a lot of literature on this, most of which is irrelevant to my comparison of the MIE approach with my abductive approach.

10.1 Matched Interlevel Experiments

My version of a science-first philosophy of science began with the idea that the sodium hypothesis is a compositional hypothesis that explains the initial inward current in a voltage-clamped axon. I then examined how Hodgkin and Huxley supported this hypothesis. A central observation of this case is that they manipulated an axon by voltage clamping it, then measured the current in that axon, and then postulated an unmeasured flux of sodium ions. They performed an intralevel experiment. Moreover, they postulated the flux of sodium because that would compositionally explain the initial inward current. Reflecting on this case, one can see the affinities between it and the interpretation of any number of other experiments in the cognitive science literature. One can present a native speaker

[1] See, for example, Harinen (2018), Krickel (2018b), and Prychitko (2021).

of a language with a string of words and measure some response, and then venture some hypothesis regarding a putative mental grammar that does, or does not, generate that string. One can have an infant observe some sequence of events and measure some response, and then venture some hypothesis regarding the infant's cognitive endowment.

With my science-first approach – with my philosophical preconception – I have found scientists using intralevel experiments to confirm compositional hypotheses. However, CG&P have a different preconception. CG&P propose that "philosophers should be guided in their thinking about constitutive relevance by attending to the *interlevel experiments* scientists use to test whether something is a component in a mechanism" (Craver et al., 2021, p. 8809, italics in original). Maybe this is a reasonable proposal, but it clearly has a limitation. By taking interlevel experiments as a guide, the CG&P approach overlooks the intralevel cases I have been writing about. In this chapter, I take up the CG&P approach and examine how scientists interpret interlevel experiments. Rather than adopt the MIE account of these experiments, I outline a singular interlevel abductive account.

Interlevel experiments might be divided into two types: top–down and bottom–up. In a top–down experiment, one manipulates an activity instance of some individual, an S Ψ-ing, and then measures an activity instance of one of its associated parts, an x_i φ_i-ing. In a bottom–up experiment, one manipulates an activity instance of some individual, an x_i φ_i-ing, and then measures an activity instance of a whole of which it is a part, an S Ψ-ing. Bottom–up experiments may also be subdivided into two types: those in which activating x_i φ_i-ing activates S Ψ-ing and those in which inhibiting x_i φ_i-ing inhibits S Ψ-ing.

Imagine a set of experiments based on an example from Craver et al. (2021). I propose these as imagined experiments, since CG&P do not cite actual experiments from the scientific literature. Let S Ψ-ing be an instance of a worm's turning when touched and let x_i φ_i-ing be an instance of the firing of an ALML neuron. Let Ψ_{in} be a touching of the worm's head and let Ψ_{out} be the worm's turning. Here are the three experiments:

1) There is a bottom–up experiment in which the worm's head is touched while the firing of an ALML neuron is inhibited, and then the worm's turning is inhibited. (This is a bottom–up inhibition experiment.)
2) There is a bottom–up experiment in which an ALML neuron is stimulated to fire and then the worm's turning is activated. (This is a bottom–up activation experiment.)

3) There is a top–down experiment in which the worm's head is touched and then the ALML neuron is found to fire and the worm's turning is activated. (This is a top–down activation experiment.)

Here, there is an imagined set of experiments, an instance of an imagined experimental practice.

How should philosophers of science interpret this imagined practice? What theory should be offered of it? CG&P propose MIE:

> (**MIE**) To establish that an entity X and its activity φ are constitutively relevant to a mechanism that Ψs, the following experimental results and matching condition are jointly sufficient:
> (CR1i) If an experiment initiates conditions Ψ_{in} while a bottom–up intervention, **I**, prevents or inhibits X's φ-ing, alterations to or prevention of Ψ's terminal conditions, Ψ_{out}, are detected.
> (CR1e) If a bottom–up intervention, **I**, stimulates X's φ-ing, Ψ's terminal conditions, Ψ_{out}, are detected.
> (CR2*) If a top–down experiment initiates conditions Ψ_{in} and detects Ψ's terminal conditions Ψ_{out}, X's φ-ing is also detected.
> (Matching) The activities φ_i activated or inhibited in bottom–up experiments (CR1i and CR1e) must be of the same kind as, and occur within quantitatively overlapping ranges with, the activities φ_i detected in top–down experiments (CR2*). (Craver et al., 2021, p. 8822)

It is reasonable to read MIE as mere joint sufficiency conditions for establishing that an x_i φ_i-ing implements an S Ψ-ing, while being entirely noncommittal about whether scientists ever use MIE. That is how the conditions are formulated. Further, it would also explain why CG&P never refer to scientific journal articles that describe scientists performing the experiments described in MIE. This would fit the description of the experiments as imagined experiments.

Then again, looking at the broader context of the CG&P paper, there are reasons to think that the conditions of MIE must be descriptive of actual scientific practice. In the Preface to Craver (2007), Craver wrote,

> My project is both descriptive and normative. My descriptive goal is to characterize the mechanistic explanations in contemporary neuroscience and the standards by which neuroscientists evaluate them. This cannot be accomplished without attention to the details of actual neuroscience. I illustrate my descriptive claims with case studies from the recent history of neuroscience. . . . For philosophers, I limit myself to the details required to demonstrate that the view corresponds to real neuroscience. (Craver, 2007, p. vii).

I am uncertain how one reconciles these claims with the idea that MIE offers mere joint sufficiency conditions for establishing compositional hypotheses. Maybe Craver intended the theory of explanation to be descriptive and normative, but the conditions of MIE to be neither. They are mere joint sufficiency conditions. Or maybe the theory of explanation is supposed to be descriptive and normative, but the conditions of MIE are meant to be merely prescriptive. Or maybe Craver thinks that MIE provides a set of jointly sufficient conditions *that scientists actually use.*

Fortunately for me, my project does not depend on how one interprets Craver's comments. My goal is to understand how scientists try to confirm compositional hypotheses in the primary experimental literature. For that goal, it does not matter what goals CG&P have for MIE. I face MIE using a constructive dilemma. Either MIE is meant to describe actual practice, or it is not. If it is meant to describe actual practice, there are features of that practice that it does not capture, namely, the role of intralevel experiments. This invites an approach that does assign a role to intralevel experiments. An abductive approach can handle these cases. If MIE is not meant to describe actual scientific practice, then, again, that leaves open the door for an abductive approach that does.

10.2 Alternatives to MIE

A first step in thinking clearly about MIE is to realize that there are alternative philosophical interpretations of the imagined experimental practice. A logical empiricist might propose that the practice be interpreted as instances of HD-confirmation. Each of the experiments could be interpretated as the testing of the hypothesis that the firing of ALML neurons implements that worm's turning behavior. One might give more complicated formulations, but for simplicity I will limit myself to the Hempelian version.

Experiment 1):

If the firing of ALML neurons implements the worm's turning behavior and the worm's head is touched and the ALML firing is inhibited, then the worm's turning is inhibited.

The worm's turning is inhibited.

= = =

Therefore, the firing of ALML neurons implements the worm's turning behavior and the worm's head is touched and the ALML firing is inhibited.

Therefore, the firing of ALML neurons implements the worm's turning behavior.

Experiment 2):

If the firing of ALML neurons implements the worm's turning behavior, then the worm turns when the ALML neuron fires.

The worm turns when the ALML neuron fires.

= = =

The firing of ALML neurons implements the worm's turning behavior.

Experiment 3):

If the firing of ALML neurons implements the worm's turning behavior, then when the worm's head is touched and the ALML neuron fires, the worm's turning is activated.

When the worm's head is touched and the ALML neuron fires, the worm's turning is activated.

= = =

The firing of ALML neurons implements the worm's turning behavior.

The HD confirmation interpretation understands each experiment as essentially backing an instance of affirming the consequent. Thus, on this account, each experiment provides a measure of confirmation of the compositional hypothesis.

The point in reviewing the HD account is not to recommend it, but to make a conceptual point. There are theoretical rivals to MIE. The HD account also serves as a springboard to my abductive account. In Chapter 4, I proposed an abductive approach as a successor to HD confirmation for intralevel experiments. In Chapter 7, I showed how the abductive approach contrasts with Peirce's version of HD confirmation. Here, I will suggest how to develop a successor to HD confirmation for interlevel experiments.

To abductively confirm the compositional hypothesis that an activity instance of a firing of an ALML neuron implements an activity instance of a worm's turning, one needs this compositional hypothesis to be included in an explanans. Further, to tie the abductive inference to interlevel experimental results, one needs the experiment to generate an explanandum for the hypothesis to explain. With these assumptions in place, let us return to the three types of imagined experiments.

The first type of experiment is a bottom–up inhibition experiment wherein the worm's head is touched while the firing of an ALML neuron is inhibited, and then the worm's turning is inhibited. This experiment raises the explanation-seeking why-question, "Why did the inhibition of the ALML neuron firing after the head touch inhibit the worm's turning?" Answer: The firing of the ALML neuron implements the worm's turning, so inhibiting the firing inhibits the turning. Further, insofar as this explanation is used to support the compositional hypothesis that the firing implements the turning, there is an abductive inference.

What sets explanations like this apart from the compositional explanations of Chapters 2 and 3 is that the explanandum is interlevel, rather than intralevel. What is to be explained is a co-occurrence of an instance of a lower level inhibition of an ALML neuron and an instance of an upper level behavior of the worm, whereas in the compositional explanations considered in Chapters 2 and 3 what is to be explained is, say, an instance of a rat navigating a maze or an axon manifesting a current. Just to have a name, call these "singular interlevel explanations" and the corresponding abductive inferences "singular interlevel abductions."

It is, of course, possible that something else explains why the inhibition of the ALML neuron firing inhibits the worm's turning, but that is just a reminder that abductive inferences are defeasible and that they are often defeated by rival explanatory hypotheses. Peirce noted this feature of abduction and Harman made it a key element in his proposal that warranted abductive inference is IBE. The New Mechanists have often drawn attention to this by saying that bottom–up experiments alone are not sufficient to establish compositional relations.[2]

The foregoing sketch of a bottom–up interlevel abductive account sets out the pieces for the interlevel abductive approach. This sketch can straightforwardly extend to the other interlevel experiments. The second type of experiment is a bottom–up excitatory experiment wherein an ALML neuron fires and the worm turns. This experiment raises the explanation-seeking why-question, "Why did the firing of the ALML neuron co-occur with the worm's turning?" Answer: The firing of the ALML implements the worm's turning, so when the neuron fires, the worm turns. Insofar as this explanation is used to support the compositional hypothesis that the inhibition implements the turning, it is used in an abductive inference.

As before, there are other possible explanations, but that does not undermine the abductive approach. Indeed, as just noted above, a central issue in abductive confirmation is the exclusion of rival explanations. Here again, the explanation is not one of the familiar types of compositional

[2] See, for example, Craver (2002, p. S92f).

explanation, since the explanandum is a putative interlevel fact. So, again, it is an abductive account that is not a singular compositional abductive account. It is a singular interlevel abduction.

Finally, there is a top–down activation experiment, wherein the worm's head is touched and then the ALML neuron fires and the worm turns. This experiment raises the explanation-seeking why-question, "Why was touching the worm's head followed by the ALML neuron firing and the worm turning?" Answer: The firing of the ALML implements the worm's turning, so that when the head is touched the neuron fires and the worm turns. The rest of the abductive story applies as before. This gives us an opening into an abductive approach to interlevel experiments.

10.3 Understanding Actual Interlevel Experiments

An ambitious critic of the MIE account might try to show that it has some fatal flaw that prevents it from accounting for any actual scientific practices. I am not such an ambitious critic. I am not even a critic in the sense that I will not argue that the account is mistaken. As far as I am concerned, there may exist cases of actual scientific reasoning in which the MIE account applies. Or MIE might have some other virtue, aside from characterizing actual scientific practice. My more modest goal is to draw attention to a feature of MIE that apparently limits the range of cases to which MIE is applicable: MIE has no role for intralevel experiments. There is no clause in MIE that refers to intralevel experiments. Indeed, insofar as the conditions for MIE are supposed to be jointly sufficient for establishing a compositional relation, this implies that the use of intralevel experiments is otiose. As against that, scientists sometimes combine interlevel experiments with intralevel experiments to confirm compositional hypotheses. Let me review two illustrations.

10.3.1 The Experience of "Flutter"

Consider a set of experiments originally noted by Kaplan (2012) to illustrate the scientific use of the Craver (2007) mutual manipulability account.[3] In these experiments, let S Ψ-ing be an experience of tactile

[3] Kaplan seems to think that the manipulability approach is used by scientists: "To motivate the mutual manipulability account, consider how neuroscientists typically go about determining components in neural mechanisms underlying cognitive capacities" (Kaplan, 2012, p. 557). Of course, CG&P may, or may not, agree with Kaplan's assessment.

"flutter." This is an experience – an activity instance – that is experimentally produced by the application of rapid sinusoidal indentations to the skin of the hand. Importantly, we should distinguish between the experience of flutter and the process of perceiving the flutter. We first touched on this experience/perception distinction in Chapter 2. The perception might begin with, say, compression of receptor cells in the hand, then continue to some point in the cortex; the experience, however, is a cortical event. In this case, the scientific focus is on the experience of flutter, rather than the perception of flutter. Further, we should also draw a contrast between perceptual behavior and perceptual experience.[4] The behavior will be triggered by the movements of a probe on the surface of the skin and may end with a bodily response, such as a verbal report or key press. By contrast, the experience begins a short time after the impression on the hand – after nerve impulses travel from the hand to the brain – and may, or may not, end before the muscular movements of the verbal report or key press. The experience is a cortical event that is typically triggered by the movement of the probe and that, in turn, may trigger a verbal report or key press. We might relate experience, perception, and perceptual behavior by saying that a behavior is the most temporally extended activity instance, the perception the next most temporally extended activity instance, and the experience the least temporally extended activity instance.

In one experiment from Talbot et al. (1968), scientists stimulated the skin of monkey participants. Let this stimulation be a Ψ_{in}. Let the participant's experience of flutter be the S Ψ-ing. The scientists then recorded activity instances in a subset of S1 neurons – so-called quickly adapting (QA) cells – that responded with action potential trains that matched the frequency of the vibrating stimulus. Let the firings of these QA neurons be the x_i's φ_i-ing. Thus, this was a top–down activation experiment. Scientists concluded that instances of the firing of QA neurons implement instances of the experience of flutter.

In a second experiment cited by Kaplan, Romo et al. (1998) used microstimulation in area 3b of S1 to induce an experience of flutter in monkeys. Reading the details of the experiment closely, the stimulus was not applied to single QA-cells, but to parts of columns:

> As S1 is organized into modules of neurons with similar functional properties (columns), we aimed to drive a cluster of 3b neurons—mostly of the

[4] This distinction is often ignored or denied in the embodied cognition literature.

QA type, hopefully—in a way that matched the periodic responses seen with mechanical stimuli. (Romo et al., 1998, p. 388)

Here, the x_i's φ_i-ing was an activity instance of a column. The response in this case was for the monkey to press one of two buttons indicating which of two stimuli had a higher frequency.

Although Kaplan does not discuss the introductory section of Talbot et al. (1968), the section is important. Talbot et al. begin by announcing that their paper seeks to combine two experimental designs: single-unit analysis and psychophysics. Single-unit analysis involves peripheral stimulation followed by measurements of single units or cells. As Talbot et al. put it: "Electrophysiological studies, particularly with the method of single-unit analysis, can now provide precise measures of the neural encoding in first-order nerve fibers of the parameters of peripheral stimuli, and of the successive relay and transformation of that neural replication from periphery to cerebral cortex" (Talbot et al., 1968, p. 301). One of the limitations of these methods, however, is that they "have so far provided little understanding of those cerebral mechanisms which, operating upon the transformed replication of the peripheral event in the primary receiving areas of the cerebral cortex, are thought to lead to subjective sensory experience and its overt behavioral counterparts" (Talbot et al., 1968, p. 301). Note that Talbot et al. here embrace the experience/behavior distinction: "Psychophysical studies ... seek to establish lawful relations between those experiences and certain physical aspects of the stimuli which evoke them" (Talbot et al., 1968, p. 301). As repeatedly noted, psychophysical experiments are intralevel experiments.[5] There were many illustrations of this in the case study of the biological basis of the Hermann grid illusion. This last comment is especially important. A major implication of Talbot et al.'s introduction is that they take a combination of intralevel and interlevel experiments to be more revealing than top–down experiments alone. Those who are guided by the idea that scientists use interlevel experiments to confirm compositional hypotheses risk overlooking this contention.

MIE clearly has no clause referring to a role for intralevel experiments. Further, by formulating MIE as a set of jointly sufficient conditions, the

[5] In an intralevel experiment from Talbot et al. (1968) that Kaplan does not discuss, the scientists stimulated the skin of human subjects and measured the subjects' reports of what they perceived. Let these reports be Ψ^*_{out}'s, to distinguish them from the monkey key presses Ψ_{out}. In ignoring this example, Kaplan again risks cherry picking his examples, selecting ones that are favorable to his case, but excluding those that are not.

implication is that the use of intralevel experiments is unmotivated. The very concept of "joint sufficiency" implies that the conditions are sufficient without the addition of anything more. This interpretation might be reinforced by comments from Prychitko (2021), which presents a precursor to MIE. Prychitko argues that top–down experiments are not sufficient to establish "constitutive relevance" (more is needed?) and bottom–up experiments are not sufficient to establish "constitutive relevance" (more is needed?), but the combination of top–down and bottom–up experiments is sufficient to establish constitutive relevance (nothing more is needed?).[6] But, if the combination of top–down and bottom–up experiments is jointly sufficient and nothing more is needed, then why did Talbot et al. use the intralevel experiment? The MIE account assigns no role to interlevel experiments, and the "joint sufficiency" reasoning suggests that an intralevel experiment is unnecessary. The implication here is that scientists must be using something other than MIE in this case.

In an earlier iteration of the foregoing argument, one reply was that the scientists did not use the intralevel experiment. Instead, the scientists only used a conclusion or assumption based on the experiment. This is an inadequate reply. In the first place, it is entirely *ad hoc*. My proposal concerns the history and philosophy of science; it is concerned with actual historical events. What historical basis is there for saying that Talbot et al. (1968) used a conclusion or assumption based on the experiment, but not the intralevel experiment? Further, what historical basis is there for saying that the Talbot et al. (1968) used the interlevel experiment, but not the intralevel experiment? Second, the reply is pointless. Even if Talbot, et al., only used a conclusion or assumption based on the experiment, rather than the experiment, MIE still has no place for a conclusion or assumption based on an intralevel experiment. Even the *ad hoc* interpretation of Talbot et al. (1968) does not enable the MIE account to correctly characterize the historical episode.

To step back a bit, the New Mechanist framing of the issue in terms of what experiments are, or are not, sufficient can be misleading. When arguing that top–down and bottom–up experiments alone are not sufficient, Craver and Prychitko point out that while it might be the case that x_i φ_i-ing implements S Ψ-ing, there are rival hypotheses about what might be going on.[7] What they look to be saying is that the interlevel experiments are defeasible and that they can be defeated by rival hypotheses. Once the

[6] Prychitko's discussion is similar to Craver's: Craver (2007, p. 154f).

[7] See Craver (2007, chapter 4) and Prychitko (2021, p. 1832).

situation is framed in terms of defeasibility by rival hypotheses, however, it is easy to see an opening for the abductive approach. A top–down interlevel experiment may provide a defeasible reason to think that x_i φ_i-ing implements S Ψ-ing; a bottom–up interlevel experiment may provide a defeasible reason to think that x_i φ_i-ing implements S Ψ-ing. Most importantly for the compositional abductive approach, an intralevel experiment may also provide a defeasible reason to think that x_i φ_i-ing implements S. What might defeat these pieces of evidence? A rival hypothesis. This framing reveals the relevance of an abductive approach. Combinations of interlevel and intralevel experiments involve combinations of distinct sets of abductive inferences. The abductive approach, thus, correctly describes the work by Talbot et al. (1968), whereas MIE does not.

10.3.2 Motion Perception and MT

Prychitko (2021) offers three experiments regarding motion perception in monkeys that *prima facie* fit the MIE account. Prychitko proposes that Newsome and Paré (1988) offers a bottom–up inhibitory experiment in which lesions to portions of MT impair motion detection. In these experiments, Prychitko's idea appears to be that MT cell activity, x_i φ_i-ing, implements a monkey's perceiving some motion, S Ψ-ing. Salzman et al. (1990) provides a bottom–up excitatory experiment in which stimulation of directionally tuned MT cells facilitates motion detection. Finally, Britten et al. (1992) offers a top–down experiment that measures activity in single MT cells when monkeys perceive motion. One attractive feature of these experiments is that they involve the same method for measuring motion perception thresholds. In these experiments, the input stimulus, Ψ_{in}, consisted of a stream of randomly positioned dots in which the percentage of dots moving together could be varied. The more dots that moved together, the stronger the movement signal. Further, an animal's response, Ψ_{out}, was to saccade to one of two targets representing the direction of motion, either up or down. Consider each of these papers in turn.

In their top–down experiment, Newsome and Paré determined the monkeys' motion sensitivity thresholds by varying the percentage of dots moving together. In their study, two monkeys were trained to reliably discriminate up from down motion with roughly 1.8% correlation. The monkeys were, thus, very sensitive to motion signals. After training, ibotenic acid was injected into MT in three hemispheres (m1, m2, and w1), with the fourth being spared as a control. Lesions raised the threshold for correlation by 400–800% for the contralateral eye, whereas the threshold for correlation was unaffected in the ipsilateral eye.

In the Salzman et al. bottom–up activation study, three rhesus monkeys were trained to discriminate the direction of motion in the random dot display. After insertion of stimulating and recording electrodes, the display was positioned in the monkey's visual field to activate MT cells primarily in a single cortical column. The display could then produce motion in the column's preferred direction or in the null direction. During the task, the column was randomly stimulated. As expected, when the direction of motion in the display was in the column's preferred direction, the stimulation enhanced the monkeys' likelihood of correctly perceiving the direction of motion. The stimulation biased the monkeys' perception.

The third study, the top–down experiment in Britten et al. (1992), showed monkeys the motion display and then measured responses in single MT neurons. Britten et al. were especially concerned to emphasize the role of single neurons. So, they write, "We compared the ability of psychophysical observers and *single cortical neurons* to discriminate weak motion signals in a stochastic visual display" (Britten et al., 1992, p. 4745, italics added) and "In fact the responses of *single neurons* typically provided a satisfactory account of both absolute psychophysical threshold and the shape of the psychometric function relating performance to the strength of the motion signal" (Britten et al., 1992, p. 4745, italics added).

Although Prychitko does not mention it, intralevel experiments play a role here. In each experiment, the monkeys had to be trained to signal the direction they perceived. This training involved presenting the dot pattern, Ψ_{in}, and measuring the monkey's response, Ψ_{out}. The interlevel experiments would not make sense without this information regarding the training. As Talbot et al. might suggest, the combination of interlevel and intralevel experiments can be more revealing than interlevel experiments alone. Philosophers of science guided only by interlevel experiments risk missing this point.

10.4 The Manipulationist "Picture" of Confirmation

On its face, one might think that the manipulationist New Mechanists can easily modify MIE so that it describes the scientific use of combinations of intralevel and interlevel experiments in the confirmation of compositional hypotheses. Just claim that MIE can use intralevel and interlevel experiments. I have given one reason to think that the manipulationist New Mechanists might resist this approach: they think about MIE in terms of joint sufficiency conditions. Now I want to give another reason to think that the MIE approach is resistant to this move. Incorporating a clause that includes a role for intralevel experiment would constitute a deep reconsideration of the manipulationist "picture" of confirmation.

A core theoretical assumption of the manipulability approach is that to find a dependency between X and Y, one (ideally) manipulates X and measures changes in Y. In 2007, Craver proposed that "Constitutive relevance is symmetrical in a way that etiological (that is, causal) relevance typically is not" (Craver, 2007, p. 153) and that "all constitutive dependency relationships are bidirectional" (Craver, 2007, p. 153). This is an ontological presupposition noted in Chapter 2.[8] Although the point is rarely noted in the literature, Craver takes the ontological presupposition to ground his epistemological view of mutual manipulability. He claims that this bidirectionality is "the core reason why constitutive relevance should be understood in terms of mutual manipulability" (Craver, 2007, p. 153). In other words, in 2007, Craver's idea was that, in compositional cases, X ontologically depends on Y and Y ontologically depends on X, so to confirm this one manipulates X and measures a change in Y and manipulates Y and measures a change in X. The picture does not have a place for manipulating X and measuring a change in Y, and then hypothesizing that U compositionally depends on V. More concretely, the theory does not have a place for stimulating a monkey's hand and measuring a saccade, and then postulating that a perception of flutter compositionally depends on the firing of QA cells.

MIE preserves much the same picture of how to epistemically determine dependencies. On the MIE picture, Z depends on Y and Y depends on X. Or X causes Y and Y causes Z. This is an ontological presupposition. Thus, the MIE methodological approach is to have an experiment manipulating X and measuring a change in Y and two more experiments manipulating Y and measuring changes in Z. Each manipulation-measurement pair is meant to confirm a dependency. Here is another way of making the point. If one wants a set of experiments that shows that X causes Y which causes Z, one needs an experiment that links X and Y and another two (activating and inhibiting) that link Y and Z. It will not

[8] To repeat a point from Chapter 2, Schindler (2013) has an apt critique of this presupposition. I have heard it claimed that everyone has abandoned this assumption so that attending to this is a misdirection. Yet, Schindler (2013) reports that, in personal correspondence, Craver does not abandon the symmetry assumption. One reason for Craver not to abandon the symmetry assumption is that it plays an important theoretical role in the mutual manipulability approach. Moreover, Krickel (2024, p. 14) seems to endorse the symmetry assumption.

do to have one experiment that links X and Y and another two (activating and inhibiting) that link Y* and Z, where Y ≠ Y*. It will not do to have one experiment that links touching the worm on the head and the firing of an ALML neuron and another two experiments linking, say, the firing of AVM neurons and the worm's turning. Confirming the chain X–Y–Z has to involve X, Y, and Z.[9] If all of that is what underlies the rationale for MIE, then what space is there in this picture for intralevel experiments?

The goal to which I have been driving in this section is that: One might think that it is a trivial matter for the manipulationist New Mechanists to simply add that intralevel experiments have a role to play. And as a comment about scientific practices that would seem to be the right response. There is abundant evidence that intralevel experiments are combined with interlevel experiments to confirm compositional hypotheses. The difficulty lies in reconciling some role for intralevel experiments within a manipulationist theoretical approach in which dependencies between X and Y are confirmed by a manipulation of X and a measurement of Y. In practice, scientists manipulate X and measure a change in Y, and then hypothesize that U compositionally depends on V. Scientists do things such as stimulating a monkey's hand and measuring a saccade, and then hypothesize an experience of flutter that compositionally depends on the firing of QA cells. The challenge is saying how the admission of a role for intralevel experiments does not amount to a sweeping abandonment of the manipulationist picture about how to confirm dependencies.

10.5 MIE as Something Other than an Account of Scientific Practice

Over the years, numerous reviewers have proposed that manipulationist accounts, such as MIE, should not be interpreted as a theory of actual scientific practice. MIE has other goals, so it is unfair to criticize it on that ground. In this section, I will review some of these proposals for what MIE is about. While I think that these are dubious readings of CG&P's aspirations for MIE, I do not mean to insist on that. I am officially noncommittal about what CG&P think about this. Instead, I want to emphasize, first, that if one thinks that MIE does not provide an account of how scientists confirm compositional hypotheses, then this clears the

[9] In an apparent nod to actual scientific practice, however, CG&P do not require that Y be literally the same in the X–Y experiment and in the activating Y–Z experiment. Instead, they "must be of the same kind as, and occur within quantitatively overlapping ranges."

field for an abductive account. Second, I want to point out how this interpretation of MIE is likely to be mere temporizing. Historians and philosophers of science should not ignore how scientists actually confirm compositional hypotheses.

MIE only claims that interlevel experiments are involved in the confirmation of compositional hypotheses. This seems to me such a dubious reading of the literature that I hesitate to address it, but here goes. As I read the texts, Craver (2002) and Craver (2007, pp. 144–152) describe the scientific practice of using interlevel experiments. By contrast, Craver (2007, pp. 152–160) evidently goes beyond a descriptive claim about scientific practice. Craver offers mutual manipulability as a theory of what underlies those practices. He is offering a theory of confirmation that is an alternative to, say, HD confirmation. If he is not offering such a theory, then what is he doing with the concept of mutual manipulability? Further, there is what is "mutual" that goes beyond the bare claim that interlevel experiments are involved in the confirmation of compositional hypotheses. Further, as I read (Craver et al., 2021), it is meant to be a clarification or technical revision of Craver's 2007 mutual manipulability account that preserves the core ideas of the mutual manipulability account. Further, as just described in Section 10.4, the "core reason" for mutual manipulability goes beyond merely requiring interlevel experiments. It involves a presupposition that the experiments are somehow describable as "mutual manipulability."

Suppose, however, that one insists that MIE holds only that interlevel experiments are involved in the confirmation of compositional hypotheses. This is a short-sighted defense of MIE. The defense can yield the conclusion that MIE is not wrong, but it leaves the MNM without a theory of those practices. In other words, the objection is that the manipulability New Mechanists do not commit an error of commission, but an error of omission, if you will. If MNM just describes a set of practices, then some historian and philosopher of science should provide a theory of the reasoning underlying those practices. The abductive approach, thus, begins to fill a gap in this take on MIE.

MIE only provides sufficiency conditions for establishing compositional relations. Unlike the foregoing interpretation, there seems to me some textual basis for this interpretation. After all, CG&P present the MIE conditions as jointly sufficient: "To establish that an entity X and its activity ϕ are constitutively relevant to a mechanism that Ψs, the following experimental results and matching condition are jointly sufficient." Suppose, therefore, that MIE is nothing more than mere joint sufficiency conditions.

Here again, this is a short-sighted defense of MIE. Concede, for the sake of argument, that MIE does constitute a set of jointly sufficient conditions for establishing a compositional hypothesis. So, MIE is not wrong and not guilty of some error of commission. Yet, why should a historian and philosopher of science care about MIE, unless it has something to do with science? Why is MIE interpreted as mere sufficiency conditions anything more than a mere philosophical curiosity? Such a view of MIE leaves open how scientists actually confirm compositional hypotheses. If one is interested in how scientists confirm compositional hypotheses, then at least the abductive approach is in the game.

MIE provides sufficiency conditions that scientists actually use in establishing compositional relations. This version of the last reply seems to me more charitable than the last. After all, claiming that the conditions are jointly sufficient does not preclude saying that the conditions are jointly sufficient conditions that scientists actually use. It would reconcile the CG&P claims about sufficiency conditions with Craver's earlier statement that "My descriptive goal is to characterize the mechanistic explanations in contemporary neuroscience and the standards by which neuroscientists evaluate them. This cannot be accomplished without attention to the details of actual neuroscience. I illustrate my descriptive claims with case studies from the recent history of neuroscience" (Craver, 2007, p. vii). Moreover, It would also make sense of why Kaplan (2012) and Prychitko (2021) refer to examples from the scientific literature: the scientific experiments are supposed to illustrate the manipulationist approach in scientific action.

Unfortunately, this interpretation simply backs into the observations in Section 10.3. The scientific examples to which Kaplan and Prychitko refer do not use MIE. The examples involve a combination of interlevel and intralevel experiments, whereas MIE assigns a role only to interlevel experiments. If MIE is supposed to provide a set of joint sufficiency conditions that scientists actually use, then it is incumbent upon those who hold this view to provide an instance in which scientists actually use them. I have no knock-down argument that there are no such cases. But, even if MIE correctly describes some cases, there are many cases where scientists plausibly use abductive reasoning.

MIE is only meant to provide norms for scientific practice. CG&P do not claim that MIE only provides a norm for scientific practice, but Craver and the New Mechanists often write about scientific norms in other contexts. So, maybe this proposal is best thought of as a possible defense of the study of MIE, rather than as CG&P's defense of the study of MIE.

In either case, the proposal is short-sighted. What is this putative norm and what reason is there to think it is a *bona fide* norm for science? Some of Craver's comments are relevant: "We can address the question of whether the norms of neuroscience are justified only when we have an idea of what the norms are and of how they can be defended. The descriptive project, in other words, is the first step in a normative project: to clarify the distinction between good explanations and bad" (Craver, 2007, p. vii). So, Craver's view in 2007 was that one cannot do the normative project without the descriptive project. Further, Craver's 2007 proposal was that philosophers of science should look at clear and uncontroversial cases of good explanations and bad explanations to distill from them the norms of good explanation. For Craver in 2007, actual practice is a guide to the norms of practice. In a different context, Craver also claims that "If one hopes to understand the norms implicit in the practice of science ... one must begin by looking at real science" (Craver, 2007, p. 235). Maybe MIE only provides norms for scientific practice rather than a description of scientific practice. Nevertheless, this is a temporizing claim since a concern for practice apparently returns as soon as one begins to take the first step in inquiring about the basis for the norms. To justify the norm, one apparently needs to appeal to actual practice.

Suppose, however, one breaks from Craver's 2007 view and allows for some *a priori* justification for MIE. The importance of actual scientific practice would still not simply disappear. If one has some *a priori* justification for MIE, one might be concerned about cases where scientists do not follow MIE. If one's *a priori* justification for MIE conflicts with actual scientific practice, one faces a substantive question. To put the matter simplistically, "Are scientists being irrational in not following the philosopher's *a priori* norms or are scientists rational and the philosopher mistaken in her *a priori* norm?" This is not a trivial question. The upshot, again, is that to say that MIE is only a norm of scientific practice, rather than a description of scientific practice, is short-sighted.

10.6 Summary

I have proposed that scientists use abductive reasoning in support of compositional hypotheses. MIE may, or may not, be taken as a rival theory. If MIE is intended to describe scientific practice, then there will be cases that are better described by the abductive approach than by the MIE approach. Maybe there are cases that are better described by the MIE approach than by the abductive approach, but that remains to be seen. If MIE is not intended to describe scientific practice, then there is one less rival to the abductive approach. Moreover, even if MIE is meant to do something other than describe actual scientific practice, one can see that a concern for actual scientific practice is still a pressing concern.

Here is how one might think about this chapter. Indeed, here is how one might think about the first two parts of this book. CG&P proposed that MIE is one method for determining what entities stand in compositional relations. In a late section of their paper, they argue that the MIE approach cannot explicate the rationale for thinking that a bottom bracket is a component in a bicycle drive train.[10] Further, in the final section of their paper, they cite other works that present what they take to be other methods of confirming compositional relations.[11]

One might read this chapter as an expansion on CG&P's comments. They proposed that the bottom bracket example, and others like it, indicate a need for an alternative to MIE. I have proposed that combinations of intralevel and interlevel experiments used in science also indicate a need for an alternative to MIE as a candidate description of scientific practice. My case for an alternative to MIE is not their case for an alternative to MIE. Further, rather than simply posing a problem, I have also proposed a solution: the scientific practice of combining intralevel and interlevel experiments is a matter of combining compositional and interlevel abductive inferences. The first two parts of this book have been dedicated to providing an outline of how compositional abductive inferences work in science. Chapter 6 broached the idea of singular interlevel abductive inferences, whereas this chapter fleshes it out a bit more.

[10] Craver et al. (2021, p. 8824). [11] Craver et al. (2021, p. 8826).

CHAPTER 11

Conclusion

The foundation for this book has been the granular examination of a scientific practice, namely, the interpretation of experimental results. Hodgkin and Huxley's development of their theory of the action potential was the central example here. This science-first methodology draws attention to any number of underappreciated features of scientific practice. First of all, it guides philosophers of science in the direction of greater explanatory pluralism. It illustrates that scientists offer compositional explanations of individuals, their property instances, and their activity instances. It illustrates that scientists offer explanations of the rates of activity instances. Close attention to scientific detail draws attention to enabling explanations of why individuals engage in the activity instances they do in terms of the property instances they bear.

Collectively, these examples also begin to redress inattention in the philosophy of science to a large class of explanations: singular explanations. These are explanations of spatiotemporal particulars. The examples also draw a connection between singular explanations, experimental data, and experimental results. Experimental results and their singular explanation play a foundational role in scientific research.

Once close attention to scientific reasoning reveals these many practices, the historian and philosopher of science will want a theory of them. At least part of the story, for some explanations, is that they are representations of compositional ontological dependencies between things in the world. This account captures many instances of the compositional explanations of individuals, their property instances, and their activity instances. It also serves as a model for enabling explanations, explanations of the activity instances of individuals in terms of the property instances of those individuals, provided one accepts that enabling is a noncausal, noncompositional determination relation between a property instance of an individual and an activity instance of that individual.

Attention to scientific journal articles reveals that scientists often engage in hypothetical reasoning. Curtis and Cole reasoned that, if the action potential difference recorded between electrode A and electrode B is entirely due to the membrane action potential, then the difference should be the same whether the potential propagates from A to B or from B to A, the difference is the same, so the difference is entirely due to the membrane action potential. Generations of philosophers of science would have interpreted reasoning like this in terms of HD confirmation. The theory of compositional abduction offers an alternative interpretation – an alternative theory – of what scientists are sometimes thinking. The abductive alternative avoids the "tacking problems" that have beset HD confirmation and offers a diagnosis of HD confirmation's susceptibility to the tacking problems.

Scientists often seem to think that the fact that what is mentioned in some hypothesis H explains something mentioned in some evidence E provides some defeasible reason to believe that the hypothesis H is true. Why would scientists think this? The theory of singular compositional abduction provides an answer. The things mentioned in H are among the things that ontologically determine that the things mentioned in E are as they are. The influx of sodium ions makes the initial inward current what it is. The scientific thought is that if the things mentioned in H were not as they are represented, then the things mentioned in E would not be as they are represented. But it is things being as represented in H that makes H true. The connection between H explaining and H being true is clear. Moreover, one can see why scientists so often care about alternative explanations or rival hypotheses. If it is the things mentioned in H*, rather than the things mentioned in H, that explain the things mentioned in E, then this defeats the explanatory value of H. The theory of compositional abduction provides an account of defeasible scientific reasoning.

The theory of singular compositional abduction also complements three nearby projects, namely, the study of Peircean abduction, the study of IBE understood as warranted abduction, and hypothesis confirmation by mutual manipulation. Peirce and Neo-Peirceans have thought that scientists use abduction to introduce hypotheses (as pursuit-worthy), which are then subsequently confirmed using hypothetico-deductivism. Perhaps, as Peirce thought, this captures some scientific thinking. Perhaps one can develop a sophisticated formal logic that characterizes this. Perhaps this paves the way for a unified epistemological model of scientific discovery, diagnostic reasoning, and other kinds of creative reasoning. The theory of singular compositional abduction has a different aim, namely, understanding some episodes of the scientific interpretation of experimental results. Scientists may use singular compositional abduction to introduce hypotheses or to make the case that some hypothesis is pursuit-worthy, but they also sometimes use it to confirm hypotheses. Following Peirce, compositional abductive inferences should be distinguished from "deductive" inferences of the consequences of some hypothesis and from the "inductive" inferences that some consequences of a hypothesis are true or false. The theory of singular compositional abduction goes beyond these observations to make the case that Peirce's deductive and inductive inferences may enable subsequent abductive inferences. The deductive inferences may reveal ontological dependencies between things in the world, whereas the inductive inferences may reveal that the ontological consequences in fact obtain. Abductive inferences might then show that ontological dependencies between things in the world enable confirming explanations.

The foundational difference between the theory of singular compositional abduction and IBE typically comes down to a distinction between abduction and warranted abduction. The assumption that an abductive inference is warranted brings with it the idea that alternative explanations or rival hypotheses are somehow eliminated. For some philosophical projects, one will want a theory of warranted abductive inference, rather than mere abductive inference. To support scientific realism, for example, one wants the abductive inference to the existence of an external world to be a warranted inference. For the history and philosophy of science project of understanding how a scientist interprets an experimental result, one does not want to begin with the assumption that in the interpretation of a single experiment, the scientist is warranted in that interpretation. Historians and philosophers of science should be open to the possibility that scientists do not always take the interpretation of a single experimental result to warrant the truth of some hypothesis. A more plausible approach allows that a single abductive inference might be part of a more prolonged scientific investigation that may, or may not, conclude with the scientific acceptance of a hypothesis.

Many versions of IBE propose that scientists choose among rival hypotheses in terms of, among other things, the simplicity and scope of explanations. The theory of singular compositional abduction does not propose that scientists always consider explanatory simplicity, although they sometimes may. More significantly, the theory of singular compositional abduction may complement many versions of IBE by drawing greater attention to considerations of scope. Indeed, the theory suggests how one might develop a theory of explanatory scope. The scope of some hypothesis might be the set of experimental results the hypothesis explains; scope might also include the consideration of experimental results that the hypothesis fails to explain, that are abductively disconfirmed.

Finally, the theory of singular compositional abduction complements New Mechanist "manipulability" approaches. Most obviously, the compositional abductive approach provides an account of the scientific use of intralevel experiments to confirm compositional hypotheses. These are cases of scientific reasoning that the manipulability approach has never discussed. In contrast to the attention to intralevel experiments relevant to compositional hypotheses, the manipulability approach has focused exclusively on interlevel experiments. The compositional abductive approach does not apply to the scientific use of interlevel experiments, but it does suggest an abductive approach that does: interlevel abductive inferences. One descriptive advantage of the combination of abductive inferences over the manipulability approach is that the combination of abductive inferences allows that scientists may use both singular compositional abduction and singular interlevel abduction in support of some hypothesis. This is something that the MIE version of manipulationism evidently struggles with, as the conditions of MIE are intended to be jointly sufficient.

I am confident that I have not considered the theory of singular compositional abduction from every angle. Throughout the book I have alluded to any number of "neighboring" inferences for which I have not even attempted to provide an account. There are surely any number of objections that I have not foreseen, especially as many of these proposals so strongly diverge from prevailing approaches. The science-first approach, in fact, highlights a pathway to revision: closely examine more science. This is a strength of the science-first approach. Indeed, I must admit that there are some rather obvious scientific inferences that resist both the compositional and interlevel abduction accounts. Some of these inferences are based on fMRI.

The gist of the problem begins with the observation that, in a typical fMRI study, one presents a participant with, say, some visual stimulus and then measures the change in blood oxygenation level in some region of interest. This is a top-down experiment, so it does not fit the compositional abductive analysis. Further, it is assumed that the level of blood oxygenation in the region of interest does not compositionally ontologically determine a visual experience or visual perception. (One might say that the level of blood oxygenation in the region of interest is not constitutively relevant or mechanistically constitutively relevant to the visual perception.) Instead, the level of blood oxygenation in some region is "coupled" to some sort of neural activity that does compositionally ontologically determine a visual perception. What one apparently needs of an abductive theory of the scientific interpretation of the fMRI experiments is some theory of this "coupling." Clearly, the theory of abductive reasoning developed here is far from the final word.

References

Aidley, D. J. (1998). *The Physiology of Excitable Cells*. Cambridge University Press.

Aizawa, K., & Gillett, C. (2019). Defending pluralism about compositional explanations. *Studies in History and Philosophy of Science Part C: Studies in History and Philosophy of Biological and Biomedical Sciences*, 78.

Aizawa, K., & Headley, D. (forthcoming). Scientific Constitutive Abduction.

Aliseda, A. (2005). *Abductive Reasoning* (Vol. 330). Springer.

Anderson, D. R. (1986). The evolution of Peirce's concept of abduction. *Transactions of the Charles S. Peirce Society*, 22(2), 145–164.

Baker, A., Tang, Y., & Turner, M. (2003). Percentage decline in masters superathlete track and field performance with aging. *Experimental Aging Research*, 29(1), 47–65.

Barlow, H. B. (1953). Summation and inhibition in the frog's retina. *The Journal of Physiology*, 119(1), 69.

Barnes, E. (1995). Inference to the loveliest explanation. *Synthese*, 103(2), 251–277.

Baumgartner, G. (1960). Indirekte Grössenbestimmung der rezeptiven Felder der Retina beim Menschen mittels der Hermannschen Gittertäuschung [Indirect determination of the size of the receptive fields of the human retina using the Hermann grating illusion]. *Pflüger's Archiv für die gesamte Physiologie des Menschen und der Tiere*, 272(1), 21–22.

Baumgartner, M., & Casini, L. (2017). An abductive theory of constitution. *Philosophy of Science*, 84(2), 214–233.

Baumgartner, M., & Gebharter, A. (2016). Constitutive relevance, mutual manipulability, and fat-handedness. *The British Journal for the Philosophy of Science*, 67(3), 731–756.

Bechtel, W. (2006). *Discovering Cell Mechanisms: The Creation of Modern Cell Biology*. Cambridge University Press.

(2008). *Mental Mechanisms: Philosophical Perspectives on Cognitive Neuroscience*. Lawrence Erlbaum.

Bechtel, W., & Abrahamsen, A. (2005). Explanation: A mechanist alternative. *Studies in History and Philosophy of Science Part C: Studies in History and Philosophy of Biological and Biomedical Sciences*, 36(2), 421–441.

Bechtel, W., & Richardson, R. C. (2010). *Discovering Complexity: Decomposition and Localization as Strategies in Scientific Research*. MIT Press.

Beebe, J. R. (2009). The abductivist reply to skepticism. *Philosophy and Phenomenological Research*, 79(3), 605–636.

Bellucci, F., & Pietarinen, A.-V. (2022). Peirce's abduction. In L. Magnani (Ed.), *Handbook of Abductive Cognition* (pp. 1–14). Springer International Publishing. https://doi.org/10.1007/978-3-030-68436-5_7-1

Bennett, K. (2017). *Making Things Up*. Oxford University Press.

Betz, G. (2013). Revamping hypothetico-deductivism: A dialectic account of confirmation. *Erkenntnis*, 78(5), 991–1009.

Bickle, J. (2023). Hodgkin's and Huxley's own assessments of their "quantitative description" of nerve membrane current. *History and Philosophy of the Life Sciences*, 45.

Bird, A. (2005). Abductive knowledge and Holmesian inference. In T. S. Gendler & J. Hawthorne (Eds.), *Oxford Studies in Epistemology* (Vol. 1, pp. 1–31). Oxford University Press.

(2020). Scientific realism and three problems for inference to the best explanation. In W. Gonzalez (Ed.), *New Approaches to Scientific Realism* (pp. 48–67). De Gruyter.

(2022). *Knowing Science*. Oxford University Press.

Blodgett, H. C. (1929). The effect of the introduction of reward upon the maze performance of rats. *University of California Publications in Psychology*, 4(8), 113–134.

Bogen, J., & Woodward, J. F. (1988). Saving the phenomena. *The Philosophical Review*, 97(3), 303–352.

Bokulich, A. (2018). Representing and explaining: The eikonic conception of scientific explanation. *Philosophy of Science*, 85(5), 793–805.

Boyd, R. N. (1983). On the current status of the issue of scientific realism. In C. G. Hempel, H. Putnam, & W. K. Essler (Eds.), *Methodology, Epistemology, and Philosophy of Science: Essays in Honour of Wolfgang Stegmüller on the Occasion of His 60th Birthday, June 3rd, 1983* (pp. 45–90). Springer International Publishing. https://doi.org/10.1007/978-94-015-7676-5_3

Britten, K. H., Shadlen, M. N., Newsome, W. T., & Movshon, J. A. (1992). The analysis of visual motion: A comparison of neuronal and psychophysical performance. *Journal of Neuroscience*, 12(12), 4745–4765.

Bromberger, S. (1966). Why-questions. In R. Colodny (Ed.), *Mind and Cosmos* (pp. 86–111). University of Pittsburgh Press.

Brooks, D. S., DiFrisco, J., & Wimsatt, W. C. (2021). *Levels of Organization in the Biological Sciences*. MIT Press.

Brown, A. M. (2019). The classics updated, or an act of electrophysiological sacrilege? *The Journal of Physiology*, 597(11), 2821–2825.

(2020). *A Companion Guide to the Hodgkin-Huxley Papers*. The Physiological Society.

Cabrera, F. (2022). Inference to the best explanation: An overview. In L. Magnani (Ed.), *Handbook of Abductive Cognition* (pp. 1–34). Springer International Publishing. https://doi.org/10.1007/978-3-030-68436-5_77-1

Campos, D. G. (2011). On the distinction between Peirce's abduction and Lipton's inference to the best explanation. *Synthese*, 180(3), 419–442.

Carey, S. (2009). *The Origin of Concepts.* Oxford University Press.

Carroll, J. W. (1997). Lipton on compatible contrasts. *Analysis*, 57(3), 170–178.

Carter, J. A., & Pritchard, D. (2017). Inference to the best explanation and epistemic circularity. In K. McCain & T. Poston (Eds.), *Best Explanations: New Essays on Inference to the Best Explanation* (pp. 133–149). Oxford University Press.

Cartwright, N. (1982). When explanation leads to inference. *Philosophical Topics*, 13(1), 111–121.

Chomsky, N. (2014). *Aspects of the Theory of Syntax.* MIT Press.

Clark, M. J., & Liggins, D. (2012). Recent work on grounding. *Analysis*, 72(4), 812–823.

Correia, F., & Schnieder, B. (2012). Grounding: An opinionated introduction. In F. Correia & B. Schnieder (Eds.), *Metaphysical Grounding: Understanding the Structure of Reality* (pp. 1–36). Oxford University Press.

Craver, C. F. (2002). Interlevel experiments and multilevel mechanisms in the neuroscience of memory. *Philosophy of Science*, 69(S3), 83–97.

(2007). *Explaining the Brain.* Oxford University Press.

(2008). Physical law and mechanistic explanation in the Hodgkin and Huxley model of the action potential. *Philosophy of Science*, 75(5), 1022–1033.

(2014). The ontic account of scientific explanation. In M. I. Kaiser, O. R. Scholz, D. Plenge, & A. Hüttemann (Eds.), *Explanation in the Special Sciences* (pp. 27–52). Springer.

Craver, C. F., & Bechtel, W. (2007). Top-down causation without top-down causes. *Biology and Philosophy*, 22(4), 547–563.

Craver, C. F., & Darden, L. (2013). *In Search of Mechanisms: Discoveries Across the Life Sciences.* University of Chicago Press.

Craver, C. F., Glennan, S., & Povich, M. (2021). Constitutive relevance & mutual manipulability revisited. *Synthese*, 199(3–4), 8807–8828.

Craver, C. F., & Kaplan, D. M. (2020). Are more details better? On the norms of completeness for mechanistic explanations. *The British Journal for the Philosophy of Science*, 71(1), 287–319.

Cronin, J. (1987). *Mathematical Aspects of Hodgkin-Huxley Neural Theory.* Cambridge University Press.

Curtis, H. J., & Cole, K. S. (1940). Membrane action potentials from the squid giant axon. *Journal of Cellular and Comparative Physiology*, 15(2), 147–157.

(1942). Membrane resting and action potentials from the squid giant axon. *Journal of Cellular and Comparative Physiology*, 19(2), 135–144.

Darden, L. (1987). Viewing the history of science as compiled hindsight. *AI Magazine*, 8(2), 33.

(2017). Strategies for discovering mechanisms. In S. Glennan & P. M. Illari (Eds.), *The Routledge Handbook of Mechanisms and Mechanical Philosophy* (pp. 255–266). Routledge.

Darden, L., & Maull, N. (1977). Interfield theories. *Philosophy of Science*, 44(1), 43–64.

Davey, K. (2023). On inferring explanations and inference to the best explanation. *Episteme*, 1–18.

Day, T., & Kincaid, H. (1994). Putting inference to the best explanation in its place. *Synthese*, 98(2), 271–295.

De Lafuente, V., & Ruiz, O. (2004). The orientation dependence of the Hermann grid illusion. *Experimental Brain Research*, 154(2), 255–260.

De Palma, A., & Pareti, G. (2011). Bernstein's long path to membrane theory: Radical change and conservation in nineteenth-century German electrophysiology. *Journal of the History of the Neurosciences*, 20(4), 306–337.

De Regt, H. W. (2017). *Understanding Scientific Understanding*. Oxford University Press.

Doppelt, G. (2014). Best theory scientific realism. *European Journal for Philosophy of Science*, 4, 271–291.

Douven, I. (2017a). *Abduction*. https://plato.stanford.edu/archives/sum2017/entries/abduction/

(2017b). Inference to the best explanation: What is it? And why should we care? In K. McCain & T. Poston (Eds.), *Best Explanations: New Essays on Inference to the Best Explanation* (pp. 4–22). Oxford University Press.

(2022). *The Art of Abduction*. MIT Press.

Earman, J. (1992). *Bayes or Bust? A Critical Examination of Bayesian Confirmation Theory*. MIT Press.

Ennis, R. H. (1968). Enumerative induction and best explanation. *The Journal of Philosophy*, 65(18), 523–529.

Fann, K. T. (2012). *Peirce's Theory of Abduction*. Springer Science & Business Media.

Firestone, C., & Scholl, B. J. (2014). "Top-down" effects where none should be found: The El Greco fallacy in perception research. *Psychological Science*, 25(1), 38–46.

Fisher, A. J., Medaglia, J. D., & Jeronimus, B. F. (2018). Lack of group-to-individual generalizability is a threat to human subjects research. *Proceedings of the National Academy of Sciences of the United States of America*, 115(27), E6106–E6115.

Frankfurt, H. G. (1958). Peirce's notion of abduction. *The Journal of Philosophy*, 55(14), 593–597.

Fumerton, R. (1980). Induction and reasoning to the best explanation. *Philosophy of Science*, 47(4), 589–600.

(2017). Reasoning to the best explanation. In K. McCain & T. Poston (Eds.), *Best Explanations: New Essays on Inference to the Best Explanation* (pp. 65–79). Oxford University Press.

Gabbay, D., & Woods, J. (2005). *The Reach of Abduction: Insight and Trial*. Elsevier.

Garfinkel, A. (1982). *Forms of Explanation: Rethinking the Questions in Social Theory*. Yale Univesity Press.

Gemes, K. (1993). Hypothetico-deductivism, content, and the natural axiomatization of theories. *Philosophy of Science*, 60(3), 477–487.

Gillett, C. (2016). *Reduction and Emergence in Science and Philosophy*. Cambridge University Press.

(2021). Using compositional explanations to understand compositional levels: An integrative account. In D. Brooks, J. DiFrisco, & W. Wimsatt (Eds.), *Levels of Organization in the Biological Sciences* (p. 233). MIT Press.

Glennan, S., & Illari, P. M. (2017). *The Routledge Handbook of Mechanisms and Mechanical Philosophy*. Routledge.

Glymour, C. (1980a). *Theory and Evidence*. Princeton University Press.

(1980b). Hypothetico-deductivism is hopeless. *Philosophy of Science*, 47(2), 322–325.

Greco, J. (2014). Episteme: Knowledge and understanding. In C. A. Boyd & K. Timpe (Eds.), *Virtues and Their Vices* (pp. 285–302). Oxford University Press.

Grimes, T. R. (1993). Explanatory understanding and contrastive facts. *Philosophica*, 51(1), 21–38.

Grimm, S. R. (2014). Understanding as knowledge of causes. In F. Fairweather (Ed.), *Virtue Epistemology Naturalized: Bridges between Virtue Epistemology and Philosophy of Science* (pp. 329–345). Springer.

Hammond, C. (2014). *Cellular and Molecular Neurophysiology*. Academic Press.

Hanson, N. R. (1960). Is there a logic of scientific discovery? *Australasian Journal of Philosophy*, 38(2), 91–106.

(1972). *Patterns of Discovery: An Inquiry into the Conceptual Foundations of Science*. Cambridge University Press.

Harinen, T. (2018). Mutual manipulability and causal inbetweenness. *Synthese*, 195(1), 35–54.

Harman, G. H. (1965). The inference to the best explanation. *The Philosophical Review*, 74(1), 88–95.

(1968). Enumerative induction as inference to the best explanation. *The Journal of Philosophy*, 65(18), 529–533.

Hartline, H. K. (1938). The response of single optic nerve fibers of the vertebrate eye to illumination of the retina. *American Journal of Physiology-Legacy Content*, 121(2), 400–415.

Haugen, T. A., Solberg, P. A., Foster, C., Morán-Navarro, R., Breitschädel, F., & Hopkins, W. G. (2018). Peak age and performance progression in world-class track-and-field athletes. *International Journal of Sports Physiology and Performance*, 13(9), 1122–1129.

Heil, J. (2012). *The Universe as We Find It*. Oxford University Press.

Hempel, C. G. (1965). *Aspects of Scientific Explanation and Other Essays in the Philosophy of Science*. Free Press.

(1966). *Philosophy of Natural Science*. Prentice-Hall.

Hendry, R. F. (2006). Elements, compounds, and other chemical kinds. *Philosophy of Science*, 73(5), 864–875.

(2024). Mechanisms in chemistry. In J. L. Cordovil, G. Santos, & D. Vecchi (Eds.), *New Mechanism: Explanation, Emergence and Reduction* (pp. 139–160). Springer International Publishing Cham. https://doi.org/10.1007/978-3-031-46917-6_7

Hermann, L. (1870). Eine Erscheinung simultanen Contrastes [A phenomenon of simultaneous contrast]. *Archiv für die gesamte Physiologie des Menschen und der Tiere*, 3(1), 13–15.

Hodgkin, A. (1994). *Chance and Design: Reminiscences of Science in Peace and War*. Cambridge University Press.

Hodgkin, A. L. (1938). The subthreshold potentials in a crustacean nerve fibre. *Proceedings of the Royal Society of London. Series B-Biological Sciences*, 126(842), 87–121.

(1951). The ionic basis of electrical activity in nerve and muscle. *Biological Reviews*, 26(4), 339–409.

(1976). Chance and design in electrophysiology: An informal account of certain experiments on nerve carried out between 1934 and 1952. *The Journal of Physiology*, 263(1), 1.

Hodgkin, A. L., & Huxley, A. F. (1939). Action potentials recorded from inside a nerve fibre. *Nature*, 144(3651), 710–711.

(1945). Resting and action potentials in single nerve fibres. *The Journal of Physiology*, 104(2), 176.

(1946). Potassium leakage from an active nerve fibre. *Nature*, 158(4011), 376–377.

(1947). Potassium leakage from an active nerve fibre. *The Journal of Physiology*, 106(3), 341–367.

(1952a). Currents carried by sodium and potassium ions through the membrane of the giant axon of Loligo. *The Journal of Physiology*, 116(4), 449–472.

(1952b). The components of membrane conductance in the giant axon of Loligo. *The Journal of Physiology*, 116(4), 473–496.

(1952c). The dual effect of membrane potential on sodium conductance in the giant axon of Loligo. *The Journal of Physiology*, 116(4), 497–506.

(1952d). A quantitative description of membrane current and its application to conduction and excitation in nerve. *The Journal of Physiology*, 117(4), 500–544.

Hodgkin, A. L., Huxley, A. F., & Katz, B. (1949). Ionic currents underlying activity in the giant axon of the squid. *Archives des sciences physiologiques*, 3, 129–150.

(1952). Measurement of current-voltage relations in the membrane of the giant axon of Loligo. *The Journal of Physiology*, 116(4), 424–448.

Hodgkin, A. L., & Katz, B. (1949a). The effect of sodium ions on the electrical activity of the giant axon of the squid. *The Journal of Physiology*, 108(1), 37–77.

(1949b). The effect of temperature on the electrical activity of the giant axon of the squid. *The Journal of Physiology*, 109(1–2), 240.

Hodgkin, A. L., & Keynes, R. D. (1955). Active transport of cations in giant axons from Sepia and Loligo. *The Journal of Physiology*, 128(1), 28.

Hodgkin, A. L., & Rushton, W. A. H. (1946). The electrical constants of a crustacean nerve fibre. *Proceedings of the Royal Society of London. Series B-Biological Sciences*, 133(873), 444–479.

Hoffmann, M. (1999). Problems with Peirce's concept of abduction. *Foundations of Science*, 4(3), 271–305.

Hollings, S. C., Hopkins, W. G., & Hume, P. A. (2014). Age at peak performance of successful track & field athletes. *International Journal of Sports Science & Coaching*, 9(4), 651–661.

Hopper, A. J., Beswick-Jones, H., & Brown, A. M. (2022). A color-coded graphical guide to the Hodgkin and Huxley papers. *Advances in Physiology Education*, 46(4), 580–592.

Howson, C., & Urbach, P. (2006). *Scientific Reasoning: The Bayesian Approach*. Open Court Publishing.

Huxley, A. F. (2004). Andrew F. Huxley. In L. R. Squire (Ed.), *The History of Neuroscience in Autobiography* (Vol. 4, pp. 282–319). Society for Neuroscience.

Illari, P. M., & Williamson, J. (2012). What is a mechanism? Thinking about mechanisms across the sciences. *European Journal for Philosophy of Science*, 2(1), 119–135.

Ioannidis, S., & Psillos, S. (2022). *Mechanisms in Science*. Cambridge University Press.

Japyassú, H. F., & Laland, K. N. (2017). Extended spider cognition. *Animal Cognition*, 20(3), 375–395.

Jenkins, C. S. (2008). Romeo, René, and the reasons why: What explanation is. In *Proceedings of the Aristotelian Society* (Vol. 108, pp. 61–84). Wiley Online Library.

Kahneman, D. (2011). *Thinking, Fast and Slow*. Macmillan.

Kaiser, M. I. (2017). The components and boundaries of mechanisms. In S. Glennan & P. M. Illari (Eds.), *The Routledge Handbook of Mechanisms and Mechanical Philosophy* (pp. 116–130). Routledge.

Kandel, E. R., Schwartz, J. H., & Jessell, T. M. (2000). *Principles of Neural Science*. McGraw-Hill.

Kaplan, D. M. (2012). How to demarcate the boundaries of cognition. *Biology & Philosophy*, 27(4), 545–570.

Katz, B. (1947). The effect of electrolyte deficiency on the rate of conduction in a single nerve fibre. *The Journal of Physiology*, 106(4), 411.

Kauffman, S. A. (1970). Articulation of parts explanation in biology and the rational search for them. *Proceedings of the Biennial Meeting of the Philosophy of Science Association*, 1970, 257–272.

Keynes, R. D. (1948). The leakage of radioactive potassium from stimulated nerve. *The Journal of Physiology*, 107, 35P–36P.

(1949). The movements of radioactive ions in resting and stimulated nerve. *Archives des sciences physiologiques*, 3(2), 165–175.

Keynes, R. D., & Lewis, P. R. (1951a). The resting exchange of radioactive potassium in crab nerve. *The Journal of Physiology*, 113(1), 73–98.

(1951b). The sodium and potassium content of cephalopod nerve fibres. *The Journal of Physiology*, 114(1–2), 151–182.

Khalifa, K., Millson, J., & Risjord, M. (2017). Inference to the best explanation: Fundamentalism's failures. In K. McCain & T. Poston (Eds.), *Best Explanations: New Essays on Inference to the Best Explanation* (pp. 80–96). Oxford University Press.

Kim, J. (1994). Explanatory knowledge and metaphysical dependence. *Philosophical Issues*, 5, 51–69.

Kirchhoff, M. (2013). Extended cognition & the causal-constitutive fallacy: In search for a diachronic and dynamical conception of constitution. *Philosophy and Phenomenological Research*, 90(2), 320–360.

Krickel, B. (2018a). *The Mechanical World: The Metaphysical Commitments of the New Mechanistic Approach*. Springer.

(2018b). Saving the mutual manipulability account of constitutive relevance. *Studies in History and Philosophy of Science Part A*, 68, 58–67.

(2020). Extended cognition, the new mechanists' mutual manipulability criterion, and the challenge of trivial extendedness. *Mind & Language*, 35(4), 539–561.

(2024). Different types of mechanistic explanation and their ontological implications. In J. L. Cordovil, G. Santos, & D. Vecchi (Eds.), *New Mechanism: Explanation, Emergence and Reduction* (pp. 9–28). Springer International Publishing. https://doi.org/10.1007/978-3-031-46917-6_2

Kuffler, S. W. (1953). Discharge patterns and functional organization of mammalian retina. *Journal of Neurophysiology*, 16(1), 37–68.

Kuhn, T. S. (1977). *The Essential Tension*. University of Chicago Press.

Ladyman, J., Douven, I., Horsten, L., & Van Fraassen, B. (1997). A defence of van Fraassen's critique of abductive inference: Reply to Psillos. *The Philosophical Quarterly*, 47(188), 305–321.

Lange, M. (2022). Putting explanation back into "inference to the best explanation." *Noûs*, 56(1), 84–109.

Lenoir, T. (1986). Models and instruments in the development of electrophysiology, 1845–1912. *Historical Studies in the Physical and Biological Sciences*, 17(1), 1–54.

Levy, A. (2014). What was Hodgkin and Huxley's achievement? *The British Journal for the Philosophy of Science*, 65(3), 469–492.

Lewis, D. (2004). Causation as influence. In J. Collins, N. Hall, & L. A. Paul (Eds.), *Causation and Counterfactuals* (pp. 75–106). MIT Press.

Lipton, P. (1987). A real contrast. *Analysis*, 47(4), 207–208.

(1990). Contrastive explanation. *Royal Institute of Philosophy Supplements*, 27, 247–266.

(1991). Contrastive explanation and causal triangulation. *Philosophy of Science*, 58(4), 687–697.

(1998). The best explanation of a scientific paper. *Philosophy of Science*, 65(3), 406–410.

(2003). *Inference to the Best Explanation*. Routledge.

(2014). Inference to the best explanation. In S. Psillos & M. Curd (Eds.), *The Routledge Companion to Philosophy of Science* (2nd ed., pp. 225–234). Routledge.

Machamer, P., Darden, L., & Craver, C. (2000). Thinking about mechanisms. *Philosophy of Science*, 67(1), 1–25.

Mackonis, A. (2013). Inference to the best explanation, coherence and other explanatory virtues. *Synthese*, 190(6), 975–995.

Magnani, L. (2001). *Abduction, Reason and Science: Processes of Discovery and Explanation*. Springer Science & Business Media.

(2009). *Abductive Cognition: The Epistemological and Eco-Cognitive Dimensions of Hypothetical Reasoning* (Vol. 3). Springer.

(2017). *Abductive Structure of Scientific Creativity*. Springer.

(2023). *Handbook of Abductive Cognition*. Springer Nature.

Marmont, G. (1949). Studies on the axon membrane. *Journal of Cellular and Comparative Physiology*, 34(3), 351–382.

Maudlin, T. (2007). *The Metaphysics Within Physics*. Oxford University Press.

McAuliffe, W. H. (2015). How did abduction get confused with inference to the best explanation? *Transactions of the Charles S. Peirce Society: A Quarterly Journal in American Philosophy*, 51(3), 300–319.

McCain, K. (2014). *Evidentialism and Epistemic Justification*. Routledge.

(2015). Explanation and the nature of scientific knowledge. *Science & Education*, 24(7), 827–854.

(2019). How do explanations lead to scientific knowledge? In K. McCain & K. Kampourakis (Eds.), *What Is Scientific Knowledge?* (pp. 52–65). Routledge.

(2022). *Understanding How Science Explains the World*. Cambridge University Press.

McCain, K., & Poston, T. (2017a). *Best Explanations: New Essays on Inference to the Best Explanation*. Oxford University Press.

(2017b). Best explanations: An introduction. In K. McCain & T. Poston (Eds.), *Best Explanations: New Essays on Inference to the Best Explanation* (pp. 1–3). Oxford University Press.

McKaughan, D. J. (2008). From ugly duckling to swan: C. S. Peirce, abduction, and the pursuit of scientific theories. *Transactions of the Charles S. Peirce Society: A Quarterly Journal in American Philosophy*, 44(3), 446–468.

McMullin, E. (2013). The inference that makes science. *Zygon®*, 48(1), 143–191.

Medawar, P. B. (1963). Is the scientific paper a fraud? *The Listener*, 70(12), 377–378.

Mill, J. S. (1843). *A System of Logic*. John W. Parker.

Miller, G. A. (1956). Human memory and the storage of information. *IRE Transactions on Information Theory*, 2(3), 129–137.

Misak, C. (2017). Peirce and Ramsey: Truth, pragmatism, and inference to the best explanation. In K. McCain & T. Poston (Eds.), *Best Explanations: New*

Essays on Inference to the Best Explanation (pp. 25–38). Oxford University Press.

Mizrahi, M. (2012). Why the ultimate argument for scientific realism ultimately fails. *Studies in History and Philosophy of Science Part A*, 43(1), 132–138.

Mohammadian, M. (2021). Abduction – the context of discovery + underdetermination = inference to the best explanation. *Synthese*, 198(5), 4205–4228.

Moore, D. H. (1975). A study of age group track and field records to relate age and running speed. *Nature*, 253(5489), 264–265.

Needham, P. (2000). What is water? *Analysis*, 60(1), 13–21.

(2008). Is water a mixure? Bridging the distinction between physical and chemical properties. *Studies in History and Philosophy of Science Part A*, 39(1), 66–77.

Nernst, W. (1889). Die elektromotorische wirksamkeit der jonen. *Zeitschrift für Physikalische Chemie*, 4(1), 129–181.

Newsome, W. T., & Paré, E. B. (1988). A selective impairment of motion perception following lesions of the middle temporal visual area (MT). *Journal of Neuroscience*, 8(6), 2201–2211.

Niiniluoto, I. (1999). Defending abduction. *Philosophy of Science*, 66(S3), S436–S451.

(2018). *Truth-Seeking by Abduction* (Vol. 400). Springer.

Norton, J. D. (2021). *The Material Theory of Induction*. University of Calgary Press.

Nubiola, J. (2005). Abduction or the logic of surprise. *Semiotica*, 2005(153), 117–130.

Nyrup, R. (2015). How explanatory reasoning justifies pursuit: A Peircean view of IBE. *Philosophy of Science*, 82(5), 749–760.

Okasha, S. (2000). Van Fraassen's critique of inference to the best explanation. *Studies in History and Philosophy of Science Part A*, 31(4), 691–710.

(2002). *Philosophy of Science: A Very Short Introduction* (Vol. 67). Oxford Paperbacks.

Oleksowicz, M. (2023). A new grammar of science. *Metascience*, 32, 15–18.

Ostwald, W. (1890). Elektrische eigenschaften halbdurchlässiger scheidewände [Electrical properties of semipermeable membranes]. *Zeitschrift für Physikalische Chemie*, 6(1), 71–82.

Overton, E. (1902). Beiträge zur allgemeinen Muskel-und Nervenphysiologie [Contributions to general muscle and nerve physiology]. *Archiv für die gesamte Physiologie des Menschen und der Tiere*, 92(3), 115–280.

Park, S. (2022). The contextual theory of explanation and inference to the best explanation. *Axiomathes*, 32(Suppl. 2), 311–326.

Pearl, J. (2009). *Causality: Models, Reasoning and Inference*. Cambridge University Press.

Pearson, K. (1911). *The Grammar of Science: Physical* (3rd ed.). Adam and Charles Black.

Peirce, C. S. (1878). Deduction, induction, and hypothesis. *Popular Science Monthly*, 13, 470–482.

(1992). *The Essential Peirce: Selected Philosophical Writings (1893–1913)* (Vol. 2). Indiana University Press.

Piccinini, G. (2020). *Neurocognitive Mechanisms: Explaining Biological Cognition*. Oxford University Press.

Pitt, J. C. (2001). The dilemma of case studies: Toward a Heraclitian philosophy of science. *Perspectives on Science*, 9(4), 373–382.

Plutynski, A. (2011). Four problems of abduction: A brief history. *HOPOS: The Journal of the International Society for the History of Philosophy of Science*, 1(2), 227–248.

Prychitko, E. (2021). The causal situationist account of constitutive relevance. *Synthese*, 198(2), 1829–1843.

Psillos, S. (1996). On Van Fraassen's critique of abductive reasoning. *The Philosophical Quarterly*, 46(182), 31–47.

(2005). *Scientific Realism: How Science Tracks Truth*. Routledge.

(2007). The fine structure of inference to the best explanation. *Philosophy and Phenomenological Research*, 74(2), 441–448.

Raman, I. M., & Ferster, D. L. (2022). *The Annotated Hodgkin and Huxley: A Reader's Guide*. Princeton University Press.

Rappaport, S. (1996). Inference to the best explanation: Is it really different from Mill's methods? *Philosophy of Science*, 63(1), 65–80.

Romo, R., Hernández, A., Zainos, A., & Salinas, E. (1998). Somatosensory discrimination based on cortical microstimulation. *Nature*, 392(6674), 387–390.

Roski, S. (2021). In defence of explanatory realism. *Synthese*, 199(5–6), 14121–14141.

Rothenberg, M. A. (1950). Studies on permeability in relation to nerve function II. Ionic movements across axonal membranes. *Biochimica et Biophysica Acta*, 4, 96–114.

Ruben, D.-H. (1987). Explaining contrastive facts. *Analysis*, 47(1), 35–37.

Salmon, W. C. (1984a). *Scientific Explanation and the Causal Structure of the World*. Princeton University Press.

(1984b). Scientific explanation: Three basic conceptions. In A. Fine, M. Forbes, & L. Wessels (Eds.), *PSA: Proceedings of the Biennial Meeting of the Philosophy of Science Association* (Vol. 198, pp. 293–305). Cambridge University Press.

(2001). Explanation and confirmation: A Bayesian critique of inference to the best explanation. In G. Hon & S. S. Rakover (Eds.), *Explanation: Theoretical Approaches and Applications* (pp. 61–91). Springer.

Salzman, C. D., Britten, K. H., & Newsome, W. T. (1990). Cortical microstimulation influences perceptual judgements of motion direction. *Nature*, 346(6280), 174–177.

Schepelmann, F., Aschayeri, H., & Baumgartner, G. (1967). Die Reaktionen der "simple field" Neurone in Area 17 der Katze beim Hermann-Gitter-Kontrast [The responses of simple field neurons in area 17 of the cat to the Hermann grid contrast]. *Pflügers Archiv für die gesamte Physiologie des Menschen und der Tiere*, 294(3), R57.

Schickore, J. (2011). More thoughts on HPS: Another 20 years later. *Perspectives on Science*, 19(4), 453–481.

(2018). Scientific discovery. https://plato.stanford.edu/archives/sum2018/entries/scientific-discovery/

(2019). The structure and function of experimental control in the life sciences. *Philosophy of Science*, 86(2), 203–218.

Schiller, P. H., & Carvey, C. E. (2005). The Hermann grid illusion revisited. *Perception*, 34(11), 1375–1397.

Schindler, S. (2013). Mechanistic explanation: Asymmetry lost. In V. Karakostas & D. Dieks (Eds.), *EPSA11 Perspectives and Foundational Problems in Philosophy of Science* (pp. 81–91). Springer.

Scholl, R. (2013). Causal inference, mechanisms, and the Semmelweis case. *Studies in History and Philosophy of Science Part A*, 44(1), 66–76.

(2015). Inference to the best explanation in the catch-22: How much autonomy for Mill's method of difference? *European Journal for Philosophy of Science*, 5(1), 89–110.

Schulz, R., & Curnow, C. (1988). Peak performance and age among superathletes: Track and field, swimming, baseball, tennis, and golf. *Journal of Gerontology*, 43(5), P113–P120.

Schupbach, J. N. (2017). Inference to the best explanation, cleaned up and made respectable. In K. McCain & T. Poston (Eds.), *Best Explanations: New Essays on Inference to the Best Explanation* (pp. 39–61). Oxford University Press.

Schurz, G. (2008). Patterns of abduction. *Synthese*, 164(2), 201–234.

Semmelweis, I. (1983). *The Etiology, Concept, and Prophylaxis of Childbed Fever* (K. C. Carter, Trans.). University of Wisconsin Press.

Şerban, M. (2017). What can polysemy tell us about theories of explanation? *European Journal for Philosophy of Science*, 7(1), 41–56.

Seyfarth, E.-A. (2006). Julius Bernstein (1839–1917): Pioneer neurobiologist and biophysicist. *Biological Cybernetics*, 94(1), 2–8.

Shan, Y. (2019). Contrastivism and non-contrastivism in scientific explanation. *Philosophy Compass*, 14(8), e12613.

(2020). *Doing Integrated History and Philosophy of Science: A Case Study of the Origin of Genetics* (Vol. 320). Springer Nature.

Shepherd, G. M. (1988). *Neurobiology*. Oxford University Press.

Sherrington, C. S. (1906). Observations on the scratch-reflex in the spinal dog. *The Journal of Physiology*, 34(1–2), 1.

Shors, T. J., & Matzel, L. D. (1997). Long-term potentiation: What's learning got to do with it? *Behavioral and Brain Sciences*, 20(4), 597–614.

Skinner, B. F. (1957). *Verbal Behavior*. Appleton-Century-Crofts.

Spence, K. W., & Lippitt, R. (1946). An experimental test of the sign-gestalt theory of trial and error learning. *Journal of Experimental Psychology*, 36(6), 491.

Spillmann, L. (1971). Foveal perceptive fields in the human visual system measured with simultaneous contrast in grids and bars. *Pflügers Archiv für die gesamte Physiologie des Menschen und der Tiere*, 326(4), 281–299.

(1994). The Hermann grid illusion: A tool for studying human perceptive field organization. *Perception*, 23(6), 691–708.

(2014). Receptive fields of visual neurons: The early years. *Perception*, 43(11), 1145–1176.

Spirtes, P., Glymour, C., & Scheines, R. (2000). *Causation, Prediction, and Search*. MIT Press.

Sprenger, J. (2011). Hypothetico-deductive confirmation. *Philosophy Compass*, 6(7), 497–508.

Strevens, M. (2008). *Depth: An Account of Scientific Explanation*. Harvard University Press.

(2012). Précis of *Depth*. *Philosophy and Phenomenological Research*, 84(2), 447–460.

Suppe, F. (1998a). The structure of a scientific paper. *Philosophy of Science*, 65(3), 381–405.

(1998b). Reply to commentators. *Philosophy of Science*, 65(3), 417–424.

Talbot, W. H., Darian-Smith, I., Kornhuber, H. H., & Mountcastle, V. B. (1968). The sense of flutter-vibration: Comparison of the human capacity with response patterns of mechanoreceptive afferents from the monkey hand. *Journal of Neurophysiology*, 31(2), 301–334.

Taylor, E. (2018). Against explanatory realism. *Philosophical Studies*, 175(1), 197–219.

Teller, P. (2010). "Saving the phenomena" today. *Philosophy of Science*, 77(5), 815–826.

Temple, D. (1988). The contrast theory of why-questions. *Philosophy of Science*, 55(1), 141–151.

Thagard, P. R. (1978). The best explanation: Criteria for theory choice. *The Journal of Philosophy*, 75(2), 76–92.

Thompson, N. (2022). *Ontological Dependence*. https://doi.org/10.1093/obo/9780195396577-0203

Tolman, E. C. (1948). Cognitive maps in rats and men. *The Psychological Review*, 55(4), 189–208.

Tolman, E. C., & Honzik, C. H. (1930). Introduction and removal of reward, and maze performance in rats. *University of California Publications in Psychology*, 4, 257–275.

Van Fraassen, B. (1978). The pragmatics of explanation. *American Philosophical Quarterly*, 14(2), 143–150.

Van Fraassen, B. C. (1980). *The Scientific Image*. Oxford University Press.

Weber, M. (2004). *Philosophy of Experimental Biology*. Cambridge University Press.

(2008). Causes without mechanisms: Experimental regularities, physical laws, and neuroscientific explanation. *Philosophy of Science*, 75(5), 995–1007.

Williamson, T. (2016). Abductive philosophy. *The Philosophical Forum*, 47(3–4), 263–280.

Wilson, J. M. (2014). No work for a theory of grounding. *Inquiry*, 57(5–6), 535–579.

Wimsatt, W. C. (1994). The ontology of complex systems: Levels of organization, perspectives, and causal thickets. *Canadian Journal of Philosophy Supplementary Volume*, 20, 207–274.

(2018). Forward. In S. Glennan & P. M. Illari (Eds.), *The Routledge Handbook of Mechanisms and Mechanical Philosophy* (pp. xiv–xvi). Routledge.

Wolfe, J. M. (1984). Global factors in the Hermann grid illusion. *Perception*, 13(1), 33–40.

Woodward, J. F. (1989). Data and phenomena. *Synthese*, 79(3), 393–472.

Wright, C. D., & Van Eck, D. (2018). Ontic explanation is either ontic or explanatory, but not both. *Ergo: An Open Access Journal of Philosophy*, 5(38), 997–1029.

Young, J. Z. (1936). The structure of nerve fibres in cephalopods and crustacea. *Proceedings of the Royal Society of London. Series B-Biological Sciences*, 121(823), 319–337.

Index

NB. Many terms, such as "constitution," "data," "enabling," "interpretation," "level," and "results," have multiple uses. This index cites only the theoretical sense of the terms introduced in this book.

Printed by Integrated Books International,
United States of America